T0258231

IET TELECOMMUNICATIONS SERIES 905

ISDN
Applications in
Education
and Training

Other volumes in this series:

ISDN
Applications in Education and Training

Edited by Robin Mason
and Paul Bacsich

The Institution of Engineering and Technology

Published by The Institution of Engineering and Technology, London, United Kingdom

© 1994 The Institution of Engineering and Technology

First published 1994
Reprinted with new cover 2006

The Institution of Engineering and Technology
Michael Faraday House
Six Hills Way, Stevenage
Herts, SG1 2AY, United Kingdom

www.theiet.org

British Library Cataloguing in Publication Data
A catalogue record for this product is available from the British Library

ISBN (10 digit) 0 85296 860 4
ISBN (13 digit) 978-0-85296-860-4

Printed in the UK by Short Run Press Ltd, Exeter
Reprinted in the UK by Lightning Source UK Ltd, Milton Keynes

Contents

Preface

On February 25, 1993, Paul Bacsich and I organised a two-site colloquium under the auspices of the Institution of Electrical Engineers and the DELTA JANUS project. The subject of the colloquium was 'Videoconferencing and audiographic groupware in distance education and training.' The British Open University in Milton Keynes was linked to University College, London with videoconferencing equipment kindly provided by PictureTel. The link was, of course, made over ISDN lines, since the main focus of the day was the educational and training applications of ISDN. Half of the presentations were made from the London end and half from Milton Keynes. One hour of the colloquium was broadcast by the EUROSTEP satellite educational channel which has Europe-wide coverage. A multi-way link-up to North America was also made during the afternoon in which PictureTel presented their view of educational videoconferencing in the United States and then connected us to Columbia University for a presentation and demonstration by one of their major users. During the whole event, some of the attendees remained in Aberdeen and received the colloquium via BT's VC7000 videoconferencing equipment, managed by LIVE-NET, the London University video network.

At the Open University, delegates had been invited from many European countries to this colloquium. Many of these were attending the next day a consultative meeting, sponsored by DG XIII of the European Commission, on the use of telematic networks in education, to which this colloquium was a precursor.

In short, the colloquium was an international event to discuss, demonstrate and experience ISDN-based teleconferencing. This book represents the enthusiasm and interest in ISDN applications which were evident at the colloquium – it is not designed to be a faithful record of presentations given on the day.

Although the colloquium was already international in its range of presenters, we have extended the presentation in this book to include contributions from more countries and from a wider cross-section of uses and pedagogical attitudes. Training settings account for nearly half of the applications described and formal education the other half. Some chapters are broad-ranging surveys; some give very detailed accounts of a specific application, and others describe systems developed particularly for ISDN.

The division of chapters into three sections represents, to some extent, the current state of educational ISDN. The lack of awareness, understanding and information about ISDN in an educational context is addressed in the first section. These chapters were written specifically for the reader who wants a general introduction to the subject. Videoconferencing, by far the most 'visible' and most developed application, is covered in the second section. Australia, the United States, France, Norway and Northern Ireland provide the settings for these chapters. A range of developing uses of ISDN is represented in the third section – from audiographics systems to multimedia desktop conferencing.

Although the number of ISDN applications in education and training is, to date, relatively limited, our book does not aim to be comprehensive. There are now far more ISDN case studies, even in the area of education and training, than can be represented in one book. The knowledgeable reader will be aware that we have not included any medical applications, nor have we covered all the European Community funded ISDN education projects. Primary and secondary schools applications are not well represented, nor are the growing library and museum uses. Even the European applications which are most numerous in the book do not adequately cover the field in France, or cover uses in Germany and Portugal. These chapters are, therefore, an introduction and representative sample of what is going on in the various education and training sectors. The reader is encouraged to follow up the references given and to be alerted by the information in the book to many other applications already in progress worldwide.

INTRODUCTION TO ISDN

The three chapters of this section provide an introduction to ISDN for those who want to know what it is, what educational uses it has and what are the significant issues regarding its spread and availability.

The first chapter tackles the technical aspects of ISDN for the novice wanting a better understanding of the educational opportunity ISDN presents. A very approachable explanation of a whole range of communication terms is given by Jacobs, which contribute to an understanding of the benefits of ISDN. Building on this framework, Jacobs presents his own perspective on the strengths and limitations of ISDN for education.

The educational value of ISDN is tackled in the second chapter with an overview of three media used over ISDN: audiographics, videoconferencing and resource-based, multimedia desktop conferencing. The merits and demerits of each are described, along with predictions about their future developments. A discussion on the nature and value of interaction in the learning process follows, with reference to various teaching media, and the chapter concludes by pointing to educational contexts most appropriate for the use of ISDN.

The third chapter takes a wide view of educational networks and looks at what contributions ISDN can make. It begins with a discussion of worldwide availability of ISDN and the kind of services which were originally envisaged for it. Bacsich then gives his own views about which areas have the most potential for use of ISDN. He looks at uses both within educational institutions and for home-based students.

PictureTel's latest entry-level videoconferencing system, the Model 150, provides full-motion, full-colour video and near high-fidelity audio, and meets the latest CCITT H.320 dial-up videoconferencing standard. Features include a full pan, tilt and zoom camera, desktop keypad operation, and an optional rollabout cart. *Picture courtesy of PictureTel*

Chapter 1

ISDN: theoretical specification, practical potential

Gabriel Jacobs

Introduction

The world of computing is littered with acronyms, some of which, if the concepts they represent become widely accepted as standards, pass into everyday speech to take on a linguistic life of their own. Many people who know comparatively little about computers regularly use terms such as DOS (Disk Operating System) or CD-ROM (Compact Disc Read Only Memory), yet would be unable to say what such acronyms stand for, even perhaps that they are acronyms at all. ISDN has not yet reached that stage, although there are signs that the term is beginning to escape from its specialist boundaries. In the meantime, educators who wish fully to grasp its benefits as well as its limitations (which I shall treat once my imaginary educator-reader is armed with some basic technical information), must begin with its full form: Integrated Services Digital Network. It is a network because it uses the telephone system. Strictly speaking, the integration of its services is a synthesis of switching techniques at the telephone exchange, but it also achieves a more readily understood integration by simultaneously offering different forms of telecommunication (voice, computer-readable data, telex, facsimile) and different media (sound, images, text – that is, multimedia). Above all, it is digital as opposed to analogue.

Basics of data transmission

Digital information

The most significant distinction, then, between an ISDN and an ordinary telephone network lies in the ISDN's use of digital technology rather than the traditional analogue technology on which telephones have been based for most of this century.

The word *digital* is a further example (and we shall see others) of the way in which specialised technological vocabulary passes into everyday language without its original significance always being fully apprehended by those who then use it. *Digital* has become synonymous with *more advanced* ('incorporating the latest digital technology'), *more accurate* (digital watches), *better quality* (digital recordings). Digital technology is, certainly, all these things and more, but for a reason which can be grasped only by understanding the fundamental distinction between digital and analogue information.

Digital information is composed of recognisable steps, while analogue information is continuous. Thus, a clock with hands for hours and minutes is said to be analogue, whereas a digital clock displays numbers (digits). An analogue clock presents information (the time) in a continuous movement as the tips of its hands describe a circle. A digital clock presents the same information but in discrete increments. It may be that those increments are very small, down to a hundredth of a second or less, but they are nevertheless incremental steps, whereas the hands of most analogue clocks appear (at least) to move continuously.

It is possible to convert information from analogue to digital form (A/D conversion) and vice versa (D/A conversion), and the same example of telling the time can be used to illustrate these processes. In order to express the time after looking at an analogue clock face, the continuously varying angles of the hands must be converted to a discontinuous form as digits: five past nine, 10.30, or (like the Speaking Clock, at the third stroke) 11.15 and 30 seconds. There is no practical way of expressing the time presented on an analogue clock face except by performing this A/D conversion. The reverse process (D/A conversion) takes place when setting the time on an analogue clock after being told what time it is.

Digital computers (that is, all modern computers) work with binary arithmetic, so at their lowest level are able to handle only two digits (0 and 1). Put another way, the only thing a computer 'knows' is whether a state is Off or On. Each of these states – a 0 or 1 – is known as a *bit*, that is to say a BInary digiT. However, a computer can so rapidly handle streams of this binary data (bit streams) that it can do wondrous things, and even appear to be performing in an analogue manner.

Sound, images and textual characters can all be represented by bit streams. An audio CD, for example, achieves the representation by having a series of tiny bumps on its surface. These are read sequentially by a laser and converted to a stream of bumps and non-bumps. This bit stream is then re-converted to analogue sound (pitch, volume, texture, timbre and so on) via an amplifier and loud-speaker. A CD-ROM works in the same way, except that no D/A conversion is required because the computer to which the CD-ROM drive is attached is already a digital device. A picture can also be digitised by converting it to a series of dots (newspaper pictures used to be produced in this way), and textual characters can be digitally represented either by a matrix of cells each of which is On or Off according to the shape of the character – as in videotex or in the system of needles on the print-heads of dot-matrix printers – or encoded as a binary number as in the ASCII system (American Standard Code for Information Interchange – yet another disconnected acronym).

Bit streams containing any kind of information can be sent along a telephone line, but the telephone system was not originally designed to handle digital data. Thus, the traditional way of doing this has been to use a modem (MOdulator/DEModulator – another example). The basic principle of a modem is as follows: at the sending end, the modem converts the stream of 0s and 1s emanating (typically) from the communications port of a computer into two separate musical tones of different pitch; these tones are then sent along the telephone line to be re-converted to 0s and 1s by the receiving modem before being passed to the receiving computer. Facsimile machines work on the same basic principle.

For direct digital data transmission between computers (as opposed to dedicated facsimile transmission), the modem process has been found satisfactory for relatively small amounts of textual or numeric information. Each letter of the alphabet, number, punctuation mark and so on can be uniquely represented by seven bits in the ASCII system. Together with two or three more bits for control and other purposes, each textual character can be thought of (for the sake of simplicity) as requiring 10 bits. Thus, if the sending speed of a modem is 300 bits per second (bit/s), 30 characters per second can be transmitted, say five English words on average (if we assume that the average word has five characters followed by a space).

It is worth noting at this point that the terms *Baud rate* and *bandwidth* are often used, essentially, to mean transmission speed. Strictly speaking, both refer to frequency range rather than data rate, though for practical purposes 300 bits per second can be thought of as 300 Baud, and bandwidth as a direct indicator of data rate. The broader the bandwidth, the higher the frequency and the greater the frequency range. Cycles per second are measured in Hertz (Hz) – 1 Hz is one cycle per second. The higher the frequency, the more cycles are transmitted per second, so the more information can be carried. Consequently, even though transmission is always actually at the same speed (the speed of light), a higher-frequency transmission is said to be faster. I shall return later, in the context of ISDN itself, to the notion of frequency bandwidth.

For various reasons, the maximum transmission speed of the most advanced modems on the market is currently about 19,200 bit/s (19.2 kilobits per second, or 19.2 kbit/s), though this speed can be usually achieved only with leased telephone lines (permanent line-connections where switching at the exchange is not involved). A new generation of modems known as V.fast (think of this as Very Fast), said to be capable of 28.8 kbit/s across the ordinary public telephone network, is at the testing stage but at present works only between two identical modems, and in any event will probably not be generally released until 1994 or beyond (Wyllie, 1993). Meanwhile, the normal maximum modem speed in widespread use across the public telephone network is 9.6 kbit/s (although 14.4 kbit/s modems are beginning to be used). At this speed, the text of a 50,000-word book can be transmitted in under five minutes, but this assumes a perfectly clear line, since any spurious line noise may confuse the receiving modem. If this modem and its associated communications software do get confused, a re-send of the corrupted data can be automatically requested. This can be done in various ways, and transparently as far as the user is concerned, but checking for errors and correcting them when necessary slows down the transmission process, so that it is not unusual over standard analogue telephone lines to achieve an average throughput of only about 6 or 7 kbit/s, even when theoretically transmitting at 9.6 kbit/s or more.

Image data

A few carefully chosen words, such as the first line of a Shakespeare sonnet, may convey an enormous amount of information to the reader, yet these words can be represented by just a few hundred bits. On the other hand, digitised sound and images require a far greater number of bits even for what may appear to be less information. So, one second of CD-quality stereo sound – not enough for that line of Shakespeare – requires over 1,600,000 bits (1.6 megabits, or 1.6 Mbit). A full-screen colour image of very high quality may require close on 40 Mbit. With data sets of this size, modem transmission in most circumstances becomes impracticable, since a single reasonably high-quality image of, say, 10 megabits takes over a quarter of an hour to transmit at 9.6 kbit/s.

File size is, then, the critical factor in transmitting images across a telephone network, and it is related to two of the three elements on which a digital graphics sub-system depends, viz: writing speed, colour resolution, and spatial resolution.

Writing speed is measured in pixels per second (*pixels*, the 'dots' that go to make up a picture, are *PICture ELementS*, or *PICS ELementS* – another example), that is to say the time it takes for a given number of pixels to appear on a display monitor. This speed is sometimes quoted by manufacturers of display technology, but it has little bearing on the viability or otherwise of transmitting images across a telephone line since file size is not affected.

Colour resolution determines how many colours can be displayed, and is partly dependent on the number of bits available to describe each pixel (the pixel depth or pixel resolution). At 8-bit resolution, a maximum of 256 colours can be displayed; at 12-bit resolution, 4096 colours, and at 24-bit resolution about 16.8 million colours. It will immediately be seen by those with a working knowledge of binary arithmetic that the number of different colours that can be displayed corresponds to one higher than the highest number representable in binary format with a given number of digits; thus binary 11111111 (eight digits) is decimal 255, plus 00000000 (zero) making 256 in all. Table 1.1 shows the relationship between a selection of pixel depths and the number of colours displayed, with some examples. As a guide, the human eye can resolve around two million colours, so 24-bit images are said to be true-colour, or photo-realistic.

Pixel depth in bits	Colours	Example(s)
4	16	VGA (PC)
8	256	Super VGA (PC)
12	4096	Commodore Amiga
24	16800000	Sun SparcStation, Apple Macintosh

Table 1.1   *Digitised image colour resolution*

Spatial resolution is the number of pixels displayed. Super VGA, for example, can offer a spatial resolution of 640 pixels across by 480 pixels down. Table 1.2 shows some sample spatial resolutions, combined with typical pixel depths for the examples given, and the resulting typical size of file required to hold the data for a single image.

Example resolution	Pixel depth	Example	Typical file size (bits)
320 x 200	2	CGA	160000
320 x 200	4	EGA	320000
320 x 200	8	VGA	640000
640 x 480	8	Super VGA	2560000
640 x 400	24	Targa-24	7680000
1280 x 1024	24	SparcStation	38400000

Table 1.2   *Digitised image spatial resolution*

The majority of installed display screens work in either VGA (Virtual Graphics Array – an acronym whose full form has once again been forgotten) or Super VGA, at typical file sizes of 640 kbit and 2560 kbit per image, respectively. Thus, at 9.6 kbit/s, a Super VGA image takes around six minutes to transmit by modem (including some error correction).

It can be seen that with the use of modems, the educational possibilities of image-data transmission are severely limited. But the figures given so far are for single (still) images. When it comes to animation or motion video, the situation becomes quite hopeless. For a completely seamless motion-video effect, about 25 images (frames) per second (25 fps) are necessary. Thus one second of full-screen 25 fps video at VGA quality will in theory take up about 16 Mbit of digital data (and in practice more than this, for reasons which need not concern us here). This is very far from broadcast-quality video which requires several times that amount of data, not including sound, but it is clear that even at only VGA quality, motion video is out of the question when using modem technology. At an optimistic throughput of around 14 kbit/s, a theoretical single second of silent VGA video (16 Mbit) would still take nearly 20 minutes to transmit, while a video segment which might be used for educational purposes lasting (say) 15 seconds, would take five hours! As with the transmission of text, it is possible to reduce this time by compressing the data, but by nowhere near enough to achieve an effect of continuous motion.

The ISDN alternative

ISDN is capable of providing an end-to-end digital connection, thus obviating the need for a modem. It is also capable of transmission speeds several times that of the fastest modem. These two facts combined plainly make it more suitable for tele-educational (and all telecommunications) purposes than a modem system. But it is not a panacea. It has its technical limitations, and understanding them is paramount to making the most efficient use of it.

ISDN technical specifications

In Europe, ISDN standards are governed by the International Consultative Committee for Telephony and Telegraphy (CCITT), the recommendations dealing with ISDN being contained in its I series, the concomitant of the X and V series which have become widely known and used in the world of telematics – X.25, X.400 and so on for inter-system connection, V.21, V.22 and so on for modems (hence the real origin of the V in those V.Fast modems). The I series recommendations are to be found in Volume 3, Fascicles 7-9 of the CCITT's *Blue Book* (CCITT, 1988).

Not all European countries have accepted the *Blue Book* standard (which, in any case, the CCITT keeps amending). In Germany and France in particular, where the basics of an ISDN service were in place before the appearance of the *Blue Book*, the earlier *Red Book* standard (CCITT, 1984) was adopted. This disparity can cause minor problems with international European connections. In 1988, ETSI – the European Telecommunications

Standards Institute –was given the task of ironing out such problems within the European Community (ETSI, 1990), though as yet complete harmonisation has not been achieved, and this despite deadlines from the European Commission. But the problems are in any case not on the scale of those associated with connections between Europe on the one hand and, on the other, North America and parts of Asia (Jacobs, 1992), where (among other incompatibilities) a different transmission speed is used.

The most basic of the CCITT recommendations concern the installation of (a) digital trunk networks (which are now fully installed in most advanced European countries), (b) digital lines to customers' premises and (c) digital exchanges. These technologies represent three out of the five fundamental parts of a telephone system, the others being Network Terminating Equipment (NTE – the end of the line and some associated equipment at a customer's premises) and Terminal Equipment (TE – telephone handsets, facsimile machines, computers and so on). Details of the technology of digital switching in exchanges need not detain us long; such switching is at the heart of ISDN, but is of less importance from the educator's viewpoint than the reasons for the limits set on the transmission speed from a user's premises. The Network Terminating Equipment has a bearing on certain uses of ISDN, and of course the Terminal Equipment is of capital significance to the user, but they again deserve less detailed technical attention in the context of ISDN and education because they should not affect transmission speed – using ISDN to best advantage assumes from the start the right Terminal Equipment.

Digital lines

Digital transmission systems are qualitatively different from analogue systems. Most analogue telephone handsets – and most are analogue, whether or not they have microprocessors for functions such as number memories or last-number re-dial – are individually connected to a local exchange by a pair of twisted copper wires (a *twisted pair*). Then, once beyond the exchange and into the trunk network, many circuits share the same cable (they are thus said to be multiplexed – a multiplexer enables a single cable to carry several combined signals). Analogue trunk systems are based on 12-channel Frequency Division Multiplexing (FDM) in which, after some frequency-translation, each trunk is allocated a frequency-band slot, the 12 slots then being transmitted jointly over a cable (or by radio). This FDM system is replaced in a digital trunk network by Time Division Multiplexing (TDM).

To understand Time Division Multiplexing, it is necessary to take a step back and, initially, to think only in terms of speech. Natural speech is of course analogue, so any digital telephone system must incorporate A/D conversion. In an analogue telephone handset, transducers convert audio energy to varying electrical energy and vice versa, but for a digital transmission system the audio energy has first to be digitised. This A/D conversion is done by so-called Pulse Code Modulation (PCM) in which the sound is sampled at 8 kHz – 8000 times a second. Sound sampling is the process of digitally defining a waveform. Digitising a photograph involves approximating the information in the image to a set of co-ordinates on an imaginary grid placed over it. The finer the grid, the better the detail (though, unavoidably as we have seen, the more bits required).

Sound is digitally defined according to the same principles but, for example, spatial resolution is replaced by amplitude resolution (for practical purposes, the degree of loudness). The number of measurements taken in one second is called the *sampling rate*, and the higher this rate, the finer the resolution of the sound. To get an idea of the kind of sound quality 8 kHz sampling means in practice, consider that sampling for an audio CD is done at over 44 kHz. But telephone communications do not require such high-fidelity sound, and 8 kHz sampling produces significantly increased clarity when compared to the old analogue system. Each 8 kHz sample in a digital telephone system is represented at 8-bit resolution. Thus the transmission speed required for a single speech channel is 8 kHz x 8 bits per second = 64 kbit/s, which is the basic capacity of a European ISDN channel (its bandwidth). Several channels occupy the same cable and an 8-bit burst is transmitted in turn from each channel. This time sharing is the basis of Time Division Multiplexing. It of course happens very rapidly, so that the process is quite transparent to the user, in much the same way as packet switching is transparent.

Indeed, there is an affinity between Time Division Multiplexing and packet switching. The latter involves parcelling up data into packets, each with an identifier, transmitting them separately (and not necessarily along the same route), then re-combining them at the receiving end; thus a PAD (another example) is a Packet Assembler Disassembler. The difference, however, is that Time Division Multiplexing – as its name suggests – is time-sensitive, a fact which, as we shall see, in certain circumstances has some bearing on the limit of its transmission-speed capability.

Digital lines at 64 kbit/s to customers' premises are well-known (a CCITT recommendation) but require special installation. ISDN is able to provide two 64 kbit/s channels via an *existing* twisted pair from the local exchange, thus giving ISDN one of its great strengths: an immediate universality.

Digital exchanges

Throughout Europe, digital exchanges are gradually replacing the old Strowger exchanges which first came into service some 80 years ago and whose fundamental design has hardly changed in that time. They are based on electro-mechanical rotary switches which connect users' telephones to each other, whereas a digital exchange is controlled by a computer, the digital switching being achieved with a system known as Time Slot Interchange (TSI). Digital exchanges in the developed world are now always built so that they need little or no modification to accommodate ISDN.

Terminating and terminal equipment

The Network Terminating Equipment is divided into two in the ISDN reference model recommended by the CCITT. Network Termination 1 (NT1) is the physical end of the line at the customer's premises. Network Termination 2 (NT2) is the customer's access point, and its main functions are control of protocols and handling a Local Area Network (LAN) or an ISPBX (ISDN Private Branch Exchange). A single Network Termination (NT) can serve up to eight Terminal Equipment access points – in other words, you can attach up to eight pieces of communications equipment (TEs) to a single NT. This can obviously be a limiting factor in certain situations, but such situations are likely to be few and far between in the world of education, and the restriction is therefore, once again, of less importance than transmission speed.

As for the Terminal Equipment itself, Pulse Code Modulation A/D conversion of sound (normally speech) is carried out either in the ISDN handset or in a separate unit. For data transmission, a computer can be fitted with an ISDN communications card. Other types of specialised Terminal Equipment can also be connected to an ISDN line, the most important of which is a digital (Group 4) facsimile machine, offering superior-quality reproduction at around 12 A4 pages of text per minute.

As well as ISDN-specific Terminal Equipment, in accordance with the philosophy of ISDN as an evolutionary rather than a revolutionary system, current analogue handsets, facsimile machines and other pieces of communications equipment can be used with ISDN lines, though such equipment must first be connected to a suitable interface (a Terminal Adaptor or TA) for protocol conversion (BSI, 1990).

ISDN speed

We have seen that a single European ISDN channel is capable of a transmission speed of 64 kbit/s, over four times the speed of a fast modem. But ISDN is not limited to a single channel – channels can be combined, and there are currently two standard combination possibilities: Basic Rate ISDN and Primary Rate ISDN.

Basic Rate ISDN consists of two 64 kbit/s so-called B-channels which together give a transmission rate of 128 kbit/s. There is also a 16 kbit/s D-channel used for signalling and other control purposes, so that Basic Rate ISDN is often referred to as $2 B + D$.

Its Time Slots (TS) for the Time Division Multiplexing are repeated 8000 times a second, thus giving 2 TS x 8 bit/TS x 8000 times a second = 128 kbit/s (144 kbit/s with the D-channel).

Primary Rate ISDN (or *30 B+ D*) offers greater capacity. It has 30 channels plus a signalling channel (and a further control channel), and an overall transmission rate of 2048 kbit/s. (North America and parts of Asia run a 24-channel Primary Rate ISDN at 1554 kbit/s. Early versions of their 2-channel Basic Rate ISDN operated at 112 kbit/s – 2 x 56.)

It is possible for ISDN to deliver even higher speeds. But this takes us into the domain of Broadband ISDN. Briefly, ISDN connections can be multiplexed so that a European ISDN can use up to a 120-channel system giving a transmission rate of about 8.5 Mbit/s, and these channels can even be further multiplexed to give up to 1960 channels running at a massive 144 Mbit/s, though no broadband commercial service of this capability is currently available. Such speeds cannot normally be achieved, in any case, via twisted pairs: they require coaxial cable or fibre-optic cables. (With optical fibres, even greater speeds are obtainable: France's existing implementation of a SONET – Synchronous Optical NETwork – is an example.)

In theory, it ought to be possible to link any number of pairs of B-channels using twisted pairs into a single coherent service, but in practice there is a limit because bit streams eventually become subject to variable time delays, causing incorrect Time Slot sequencing. Here, then, is one limitation of Time Division Multiplexing.

Educational potential and limitations of ISDN

In describing some of the technical aspects of ISDN, I have dealt primarily with speech, but of course a digital network is inherently suitable for data transmission, a capability which is clearly of the utmost interest to educators. By replacing a Pulse Code Modulated speech-channel card with a data card, a 64 kbit/s data capacity is available. ISDN channels can be used separately for voice and data, and even link the user simultaneously with two locations so that, for example, a database can be interrogated and an entirely separate voice conversation carried on at the same time. But, as I have already inferred, it is also possible to use more than one channel for data. Why, then, not look to more than the two channels of Basic Rate ISDN?

Here, we come across – paradoxically – both the major limitation as well as one of the major advantages of ISDN as an evolutionary concept, and I can therefore at this point come to a suitable mini-conclusion. The plus side is that Basic Rate ISDN can be used with existing wiring at customers' premises. The minus side lies in the two-channel constraint imposed by a single twisted pair. While Primary Rate ISDN, with its 30-channel high data-transmission capability, is being used in certain educational applications such as videoconferencing, the cost of installation (and use) means that it tends to be limited to closed communities. Moreover, Primary Rate ISDN is generally perceived as a high-level technology for use only by corporate customers or large institutions. But if ISDN is to fulfil its true potential, it must become more than a system of dedicated point-to-point connections – ISDN islands as they are sometimes called. Almost everyone has, or has access to, a single twisted pair, which is one of the reasons for the popularity of facsimile. Facsimile is now a universally accepted medium because there are so many installed facsimile machines, and there are so many installed facsimile machines partly because they work with no modification to the user's telephone system. Basic Rate ISDN already has both a present and a future in education, but its widespread acceptance as an everyday standard would boost its educational usage beyond measure.

The point is well exemplified by taking an eagle's eye view of the history of personal-computing standards. The best predictor of future performance is past performance, and nowhere is this more true than in the personal-computer market. It is not that innovation or improvement is suppressed in the domain of personal computing – on the contrary – but market forces have so far determined that the fastest, the easiest to use, the most complete machine does not necessarily win the market share, and therefore the extent of its acceptance as a standard, it may in principle deserve. One has merely to take the battle between the Apple Macintosh and the IBM-compatible PC to see that this is true. Multimedia, which is now without any doubt the future of personal computing (and has clearly been so since the mid-1980s) is a far better-controlled affair on the Macintosh than on the PC, not least because the PC was not originally built for a graphical user interface such as Windows. Furthermore, the much vaunted advantage of compatibility within the PC world is a myth: the PC has spawned innumerable incompatible standards (there are, for example, over 30 different PC image formats), and third-party hardware add-ons simply cannot be relied on to work with every supposedly compatible machine. Yet the PC reigneth, in terms both of numbers of machines used world-wide and of available software and add-on hardware. Pricing has no doubt had more than a little to do with this state of affairs, and a snowball effect has been seen: the more PCs bought because of a comparatively low price tag, the more third-party suppliers (both software and hardware) have perceived a potentially lucrative market; the more third-party material available, the more PCs sold; the more PCs sold and used in business, the more sold for use in the home for compatibility with the office machine.

The same criteria will no doubt apply, either successfully as with the PC or less successfully, as so far in the case of Basic Rate ISDN. The existence of a more powerful and more versatile technology in the form of Primary Rate ISDN (which requires considerable investment on the part of the user) will not, as I see things, have any appreciable effect on the future of Basic Rate ISDN. Basic Rate ISDN will become a widely accepted norm for data transmission in education only if it becomes the generally accepted standard in business and in the home – in fact, if its progress could be boosted with an entertainment factor, its chances of success would double, at least.

The practical effect

I can now fulfil my earlier promise to treat the benefits and limitations of ISDN for the educator, and more precisely in the context of its potential as an immediately available resource, by addressing the question: to what extent does the transmission speed offered by the two-channel Basic Rate ISDN actually restrict its practical educational use?

Conveniently, each Time Slot is 8 bits wide, so the 7-bit ASCII character set plus an extra bit used (among other things) for word-processing control functions, and the 8-bit code used by, for example, executable programs, are perfectly accommodated. No A/D or D/A conversion is needed, and with digital trunk lines error rates are reduced because they are not subject to line noise. The 50,000-word book which took some five minutes to transmit with a fast modem (9.6 kbit/s), takes only around 20 seconds using Basic Rate ISDN (using the two B-channels). Text, then, presents no problems either as pre-prepared educational material or as a real-time form of interactive communication. Even with modem technology, text-only material has many worthwhile educational uses; with ISDN its scope is that much wider. But, as I have already insisted, text-only applications are not the future, nor indeed the current state of the art, and while they clearly have a place in certain educational contexts, they cannot compete with multimedia applications.

At 128 kbit/s, a Super VGA image – the rough equivalent of the 50,000-word book – can, again, be transmitted in a little over 20 seconds. A 16-colour VGA image will take only five seconds. As for motion video, the one second of VGA-quality moving images at 25 fps (16 Mbit) which took, at the very best, 20 minutes to transmit by modem, can be transmitted in under 2 minutes. That is still far from a seamless motion effect, but when one is that much closer to the goal than with modem transmission, other techniques can be applied to bring one at least within shooting distance. For both still and moving images, one can (a) reduce colour and/or spatial resolution and (b) compress the data before sending. In addition, for moving images one can opt for a frame rate lower than 25 fps, so that a perhaps jerky video effect – but nevertheless a video effect – is produced. Combinations of these techniques can be used, so that the 128 kbit/s speed still leaves room for sound and text. Each technique brings with it, however, its own restrictions – some more serious than others – which will lead me to the very heart of the problems associated with using Basic Rate ISDN for remote education.

Resolution reduction

Reducing colour resolution produces noticeably inferior (often more grainy) images, but such images may be theoretically adequate for many educational purposes. Reducing spatial resolution without reducing the dimensions of an image also produces a more grainy effect, or a 'pixelated' effect (where the pixel blocks can be distinguished) can result if a method of pixel replication has been applied to increase the size of an image without reducing its basic information content. However, spatial resolution can be reduced without loss of quality by using overall dimensions smaller than the size of the original image, typically a window occupying part of a screen. This is obviously an attractive proposition, since talking heads, for instance, are perfectly acceptable – in fact, very useful – in educational applications. On the other hand, there are occasions when a full-screen image or full-screen video are desirable.

Data compression

There is nearly always some inherent redundancy in a bit stream, in that elements are bound to be repeated. If that repetition can be encoded in a more compact form, the data will occupy less space. The simplest way of thinking about this is to take a stream of, say, 16 (decimal) numbers such as 333 44444 6666 7777, and to represent it as three threes, five fours and so on. This gives only eight elements (33 54 46 47) rather than the original 16, thus achieving a compression ratio of 2:1.

Most data compression algorithms are far more complex than this, and much higher compression ratios can be achieved. Digitised images are the most space-hungry form of information, so this is where the greatest efforts at compression have been concentrated. Sound cannot currently be satisfactorily compressed without unacceptable loss of quality at ratios much higher than 5:1 (though new methods are constantly being tried – see, for instance, Byte, 1992).

The *de jure* standards for image compression have been defined by JPEG (the Joint Photographic Expert Group) for still images, and by MPEG (the Motion Picture Expert Group) for digital video. Both groups have produced recommended sets of procedures with similar bases. In MPEG, for example, each frame is divided into 8 x 8 pixel blocks which are encoded as luminance (brightness) and chrominance (colour) values. Each block is then transformed into a series of frequencies. The resulting frequency information is readily compressible because, for example, changes in pixel intensity from one frame to the next tend to be gradual and so concentrated at lower frequencies, which means that higher frequencies can be discarded. Then a quantisation method is applied. This involves smoothing out further minor differences in frequency, therefore increasing the likelihood that adjacent frequencies will be identical. Any that are identical can be encoded in (roughly) the compressed form outlined above in the example of repeated digits. With these techniques it is possible to achieve usable compression ratios of between about 15:1 and 30:1, say 256 kbit for a photo-realistic 640 by 400-pixel frame. High transfer rates are therefore still required for digital video using the MPEG method: a single image at this resolution will obviously take about two seconds to transfer at 128 kbit/s.

A different image-compression method is one based on fractals (Barnsley, 1988; Anson, 1991). Here, a picture is again divided into pixel blocks, but they are then matched against a library of fractal shapes (shapes that retain their form no matter how many times they are magnified) represented by compact sets of numbers. This can be a long process (several minutes, even hours), but unlike other data compaction methods, decompression is not the reverse of compression, since once the relevant numbers have been found and stored, there is no need to search for the shapes again. In fact, decompression time is quite rapid (a second or so). Compression ratios of several hundred to one are achievable, and almost lossless results can be obtained if the compression ratio is not too high (a quite usable 100:1 is possible). This is because fractals offer a rich source of shapes which approximate closely to shapes in the real world, and which can be represented in roughly the same way as in standard computer graphics based on geometric shapes (a circle, square or other shape can be represented by a mathematical formula rather than describing every pixel which goes to make up the shape on the screen). A simple calculation shows that the 128 kbit/s transmission speed offered by Basic Rate ISDN is capable of delivering full-screen full-motion video at VGA quality with a compression ratio of 125:1 (our one second of VGA-quality video at 16 Mbit can be compressed at this ratio to become 128 kbit). This does not allow any room for sound or other data, but compression ratios in excess of 125:1 can be achieved with some loss in picture quality. The system does not offer compression on the fly – the compression process is far too long drawn out – so it cannot be used for video teleconferencing or other real-time interactive video applications, but it could be used for many educational applications in which compression has taken place prior to transmission (Jacobs, 1993).

For ISDN, the CCITT recommends its H.261 video data-compression standard, introduced last year after considering all available compression methods; one of the principal requirements was compression and decompression on the fly. The H.261 standard incorporates numerous extra techniques such as a system of frame difference, in which each frame in a video sequence is encoded not in its entirety but only as the differences between it and the frame which preceded it. So, if two adjacent frames consist of a static background and a moving object, only the movement need be encoded, thus saving space (this technique is well known to those who deal with Intel's DVI and certain other digital motion-video systems). However, the H.261 standard, as opposed to JPEG, MPEG and fractal transforms, requires some special video compression/decompression hardware called a codec (COder/DECoder) which can, in practice, decode (decompress) satisfactorily at between only 10 and 15 frames per second to give a rather staccato video effect. What is more, a codec represents another piece of add-on hardware, and there is always consumer resistance to add-ons.

Sound

The question of audio in conjunction with video also has obvious implications here, since digitised sound, as we have seen, is not yet as compressible as a digitised image. And the problem has been a constant thorn in the side of those working on image compression for video, not only because of overall data size but also in relation to synchronisation. This is a separate subject in itself but, briefly, the most acceptable solution at present (for both telematics and for transfer to a computer from a storage medium such as a hard disk or a CD-ROM) is to interleave the audio element with the video element in real time. Since most personal computers cannot perform more than one operation at any one time, perfect synchronisation with such interleaving is not possible, but (as with ISDN itself which, as we have seen, is time-sensitive) the human brain can fairly easily be fooled by very rapid time sharing. On the other hand, certain types of synchronisation – in particular, lip synchronisation – are more important than others. If lip sync is not perfect the results can be disastrous (think only of the film *Singing in the Rain*). In order to avoid this kind of disaster, priority is given to the audio element, since this is the element which cannot be altered without loss of significant information. If the audio and video lose synchronisation, one or more frames of the video are skipped. This can be done 'intelligently' if the controlling software is able to look ahead and judge in advance the best places to drop a frame – much as a good word processor, when putting in micro-spaces for justified text, evenly distributes them on each line instead of only towards the end of the line – but there is as yet no such predictive interleaving system which can guarantee apparently flawless performance. Rather, current interleaving systems tend to drop several frames at the same time so that the video can keep up with the audio, and the effect can be disconcerting. Of course, given a fast enough transmission speed, the problem disappears, but again we hit the technical limitations of the 128 kbit/s Basic Rate ISDN.

The educational heart of the matter

ISDN limits

ISDN is, then, quite well suited to slow-scan video, where a full-motion effect is not attempted but images nevertheless follow each other sequentially. In applications such as remote medical diagnosis (Akselsen, 1993), a screen-refresh rate of a few seconds may not be an intolerable nuisance. For example, France's ISDN service, Numéris, has been successfully used in a number of applications involving slow scanning, a noteworthy one being an EC-funded research programme (MultiMed) in which multimedia workstations in hospitalsin Rennes and Nantes are connected so that doctors can compare medical images in (more or less) real time (DGXIII-F, 1990). Other applications to be found in various European countries include tele-surveillance of premises, electronic publishing, museum guides, electronic sub-assembly manufacture, aircraft maintenance, and real-estate (Clark, 1991; TE, 1992). ISDN is also clearly suitable for educational applications involving, for example, still-image transmission on one channel, with the other channel used for voice interaction between student and teacher.

But for education and training generally, a powerful pedagogical factor comes into play when slow picture-refresh rates, and other half-baked solutions imposed by the restricted bandwidth of Basic Rate ISDN, are the norm. Northwood (1991) describes a revealing experience not unconnected with that pedagogical factor. While creating an illustrated history database for school children in the North of England, budgetary restrictions and the relatively low level of PC hardware in local schools forced him into opting for monochrome images. It seemed an unimportant constraint at the design stage, since colour was not fundamental to the application. Today's school children, however, know little other than colour media formats (television, photographs, magazines, newspapers), so that the most serious criticism of the application coming from its users at the feedback stage was the lack of colour. I can add here an experience of my own. In 1991, at the launch of Commodore's CDTV (Commodore Dynamic Total Vision, not Compact Disc TV, though that is what it is), some full-colour full-screen digital video was demonstrated, running without compression at 7 fps. The effect was naturally very jerky, but those with a professional interest in the technology were enormously impressed by the programming skill which had gone into this outstanding achievement. For those without such an interest, and without an understanding of the level of technical accomplishment, the demonstration proved only that digital video was not possible on the CDTV.

In addition to being available to all existing telephone subscribers, if ISDN is to capture a mass market, its multimedia capabilities must not fall too far short of the standards set by commercial television. Jerky video, badly synchronised audio and video, grainy or heavily pixelated images resulting from a reduction in colour and/or spatial resolution, monochrome images even if they fulfil theoretical needs, slow picture-refresh rates – all these solutions will do nothing to enhance the reputation of ISDN, its general acceptance, and therefore its wide use in education. Two seconds waiting for an image to appear on a screen may be the result of a technological marvel, but it can be an eternity for a learner staring at the process, or – worse – at nothing. Impatience is a barrier to learning, and we can do without such barriers.

ISDN potential

Such obstacles to the educational potential of Basic Rate ISDN can nevertheless be negotiated, even if this means side-stepping them rather than facing them head-on – in other words, being prepared to grasp ISDN as an educational medium and squeeze out of it as much as one can. Other contributors to this book describe the way in which they have successfully implemented ISDN educational applications, and report ways in which they have dealt with latent pitfalls, but some general principles in any event emerge from my own foregoing analysis.

Slow-scan ISDN image transmission can be implemented on one channel with a return audio (and/or textual) channel for interactivity. However, with both channels in use it can give full-screen picture quality as good as or even better than that of commercial television, with picture-refresh times of about four seconds for monochrome and about 30 for colour, and still leave room for some textual or sound interaction between teacher and student while one image is replacing another. Still images can also be left displayed on a screen while some pointing or other means of real-time commenting on the image, or some different form of interactivity, is taking place.

Furthermore, there is a vogue for interactivity in education which may not be quite as necessary to learning (or anything else) as is currently assumed. It is associated with supposedly new ideas about discovery-based and experiential learning, spontaneous thinking, the value of guesswork and so on, where the teacher is a guide rather than a dispenser of knowledge, but where a degree of interactivity is crucial. Yet learning can also take place as a one-way process, that is to say passively from the learner's standpoint. I can also return here to the launch of Commodore's CDTV, which was heavily promoted as interactive television. It has failed to capture a wide user base on this score not least because what people seem to want from television is its very passivity. If Basic Rate ISDN is used as primarily a one-way system, in which information is dispensed with a minimum of return usage, there are applications in which its capacity can be essentially doubled. I am aware that not everyone will accept a widespread passive use of ISDN in education, but it must be conceded that there are educational applications for which it is adequate, and even some for which it is the most appropriate approach.

Conclusion

When, in 1989, BT launched its ISDN-2 service (Basic Rate ISDN), the *Financial Times* claimed that the existence of two separate channels each of 64 kbit/s had suddenly opened up the possibility of multimedia communications such as desktop conferencing (FT, 1989). This claim was of course justified, but it is also reminiscent of news flashes concerning medical breakthroughs which turn out to require many more years of research and testing before they may possibly be of benefit. Sudden medical breakthroughs of the likes of smallpox vaccination or penicillin are extremely rare; far more usually, they are eventually achieved after countless small steps have been taken, so that they appear to come upon us gradually.

Progress in ISDN is in the same league – but with the crucial difference that the technology is already implemented. Basic Rate ISDN can be used right now to educational ends by anybody with a telephone, and despite the deficiencies associated with it, it still offers an attractive horizon for tele-education. The problem is that user-expectations are in advance of what the available, relatively inexpensive technology can deliver (the reverse of what is usual in the world of consumer-electronics, where the technology moves on before the market is ready for it). For example, it would be perfectly possible to produce a comparatively cheap Basic Rate ISDN videotelephone providing smooth motion video of acceptable quality, with interleaved audio – if it handled only monochrome images: colour, as we have seen, eats up bandwidth. Telecommunications companies, no doubt rightly, do not want to repeat the kind of Northwood experience I described above, yet seem content to produce colour videophones for which the quality of video is below acceptable standards. It is a paradox which typifies the apparently simple but in fact very complex relationship between demand and supply in the information-technology market. The difficulty will certainly disappear by the end of the century, when Broadband ISDN will be the norm, though it may well then be replaced by a similar problem (user-expectations of virtual reality?) and always provided that newer, faster communications technologies beginning to emerge, such as ADSL (Asymmetric Digital Subscriber Lines) or HDSL (High-rate Digital Subscriber Lines) have not by then caused a standards chaos. In the meantime, the medium-term future of current ISDN in teaching and learning depends on a cautious and well-informed but committed approach to it by educators. It *can* deliver – as the examples in this book show – and should be delivering.

Acknowledgement

I am grateful to BT for their assistance in verifying the technical accuracy and up-to-dateness (April, 1993) of the information contained in this chapter.

References

Akselsen, S., Eidsvik, A.K., and Folkow, T. (1993). Telemedicine and ISDN. *IEEE Communications Magazine*, **31**, 1.

Anson, L. and Barnsley, M. (1991). Compression technology: a new method for image reproduction using the fractal transform process. *SunWorld*, October.

Barnsley, M. and Sloan, D. (1988). A better way to compress images. *Byte*, January.

BSI (1990). *Connectors for Access to the Integrated Services Digital Network*. British Standards Institution, London.

Byte (1992). Digitally speaking. *Byte*, November.

CCITT (1984). *The Red Book*. CCITT, Paris.

CCITT (1988). *The Blue Book (I.420 Standards)*. CCITT, Paris.

Clark, P. (1991). ISDN: the multi-media enabling technology for Europe. *Proceedings of the Third IEE Conference on Telecommunications*. IEE, London.

DGXIII-F (1990). *Research and Development in Advanced Communications in Europe: RACE '90*. Commission of the European Communities, Brussels.

ETSI (1990). *ETR 010: The ETSI Basic Guide on the European Integrated Services Digital Network*. European Telecommunications Standards Institute, Valbonne.

FT (1989). A picture tells a thousand words. *Financial Times*, December 6.

Jacobs, G. (1992). Integrated Services Digital Network: state of the art in Western Europe and potential worldwide. *Proceedings of the International Conference on Trans-European Engineering Education*. UNESCO Centre for Scientific Computing, Prague.

Jacobs, G. (1993). Data compression and ISDN: possibilities and limitations for language learning. *Proceedings of the AELPL Conference on Technology and Language Learning.* Institut National des Télécommunications, Paris.

Northwood (1991). Developing and using hypermedia resources. *Interactive Multimedia,* **2**, 1.

TE (1992). ISDN application. *Telecom Europa ISDN Newsletter,* August 17.

Wyllie, J. (1993). High speed becomes standard fare. *Trend Monitor Report: Communications Industries 1992-1993.* Trend Monitor International, Portsmouth, p. 2.

Further reading

Bocker, P. *et al* (1988). *ISDN: The Integrated Services Digital Network.* Springer, Berlin.

Brewster, R. L. (1993). *ISDN Technology.* Chapman and Hall, London.

Clark, W. J., Lewis, D. E. and Lee, B. (1993). ISDN desktop conferencing with the multipoint interactive audiovisual system. *BT Technology Journal,* **11**, 1.

Fuchs, G. (1992). ISDN: the telecommunications highway for Europe after 1992. *Telecommunications Policy,* **16**, 8.

Griffiths J. (ed) (1990). *ISDN Explained: Worldwide Network and Applications Technology.* Wiley, Chichester.

Haslam, R. and Johnson, C. (1992). ISDN service and applications support in the UK. *British Telecommunications Engineering,* **11**, October.

Helgert, H. J. (1991). *Integrated Services Digital Networks: Architectures, Protocols, Standards.* Addison-Wesley.

Horn, R. W. (1992). End user requirements and solutions. *IEEE Communications Magazine,* **30**, 8.

Hoshi, T., Mori, K., Takahashi, Y., Nakayama, Y. and Ishizaki, T.B. (1992). ISDN multimedia communication and collaboration platform using advanced video workstations to support cooperative work. *IEEE Journal on Selected Areas in Communications,* **10**, 9.

Hunter, J. D. and Ellington, W. (1992). ISDN: a customer perspective. *IEEE Communications Magazine,* **30**, 8.

Kimura, H. *et al* (1993). Broad-band ISDN – Application, Networking and Management. Special issue of the *IEICE Transactions on Communications,* Japan.

Kundig, A.T. (1992). ISDN and B-ISDN: from basic technology to applications. *IFIP Transactions: C-Communication Systems,* **6**.

Lindgren B. and Jonsson, L. (1989). *Illustrated ISDN: A Painless Primer to Principles and Protocols.* Infotrans, Kalmar.

Muller, M. A. (1992). *Introduction to Digital and Data Communications.* West Publishing Co., New York.

Ronanye, J. (1990). *The Integrated Services Digital Network from Concept to Application*. 2nd edition. Pitman, London.

Rosenbaum, A. (1992). Europe looks to 93 for rapid growth in global ISDN. *Electronics*, **65**, 6.

Stallings, W. (1989). *ISDN: An introduction*. Macmillan, New York.

Thachenkary, C. S. (1993). ISDN: Six case-study assessments of a commercial implementation. *Computer Networks and ISDN Systems*, **25**, 8.

Van Binst P. and Wilkin L. (eds) (1987). Proceedings of the conference on communication and data communication. *Computer Networks and ISDN Systems*, 14, 2-5.

Chapter 2

The educational value of ISDN

Robin Mason

Introduction

The road show advertising the wonders of yet another educational revolution has set off to the usual tune of brass band, hawkers and slick sales pitches: ISDN is the mega-medium which is going to transform education, turning boring books and lectures into cost-effective, multimedia, convenient choices, easily adapted to the demands of lifelong learning, busy schedules and specialist requests. Above all else, ISDN will be interactive.

Can the educationalist rev up any enthusiasm for this latest offering? Will 'technology' really deliver this time? And what is wrong with education that it needs saving by ISDN anyway? This chapter will look closely at these questions and aim to provide those involved in education and training with a 'learners guide' to ISDN.

The chapter will survey the current applications running over ISDN, commenting on the range of teaching/learning strategies possible, and looking forward to developmental trends in software. A critical analysis will be made of the vogue for interaction in education and training, including a review of the kind of interaction which actually takes place through typical ISDN applications. Finally, the chapter will put ISDN in the context of other educational technologies and indicate those areas in which its implementation is most appropriate.

Institutional context

Current pressures in educational and training institutions which have relevance to the use of ISDN can be summarised as follows:

- financial constraints as well as a need to increase student numbers
- a dispersed population requiring specialist, up-to-date training
- a perceived demand for interactive teaching strategies
- regulations requiring equal access to educational facilities

- increasing time constraints and other commitments operating against fixed location, face-to-face education and training.

These pressures have been building for some time and have led to a world-wide expansion in various forms of distance education. Technological developments have undoubtedly made this expansion feasible: twenty years ago, distance education turned to broadcast television and radio, then to video and audio cassettes, then to computers and satellites.

Whether driven by educational demand, or pulled by technology developments, we are now very strongly in the grip of an 'interaction paradigm' for the learning experience. Technologies which offer some form of interaction are, of course, already available on analogue telephone lines and are being used: audioconferencing is an example of a real-time application, and computer conferencing and electronic mail are examples of asynchronous uses. What ISDN promises above all is better, cheaper, faster *video* interaction.

The pronounced divide between face-to-face teaching and distance teaching is simply disappearing with the host of educational technologies and institutional strategies for responding to the pressures listed earlier. Examples include:

- students at traditional campuses taking some courses via conferencing, using equipment on site to access teaching at a linked institution (videoconferencing) or even from the same institution, but in their own time (computer conferencing)
- professionals taking a course (e.g. MBA) in the workplace delivered by print, video or audiographics from a distant university, but with a local facilitator or tutor from their own organisation
- courses which combine face-to-face seminars, residential weekends or lectures with computer conferencing as the main delivery medium
- groups of students at several sites linked by audiographics with the teacher taking turns at each site to run the session.

Where once distance education and training were criticised for being over-packaged, isolating and uniform, the new interactive, technology-based learning (the concept of distance has faded away) is much vaunted for offering consumer choice, convenience and contact. As the advertising slogans present it: "better than being there!"

Some predictors say that ISDN will make distance learning the norm (Rajasingham, 1990) and face-to-face contact will have to be justified as a special case (Romiszowski, 1990).

ISDN applications

In educational terms, ISDN is a set of international standards for computer and communications technologies, which support video, audio and computer data on a telephone line. In theory, therefore, it means that multimedia material can be accessed, processed and stored through the telephone network we already have in our homes and offices. Although there are a number of companies offering this 'desktop multimedia conferencing', there are very few uses being made of these facilities for education and training as yet. This is due to the cost of the equipment/software/phone charges, the lack of penetration of ISDN, and the inevitable lag time for educationalists to develop innovative programmes. In fact, current applications are primarily modifications and enhancements of technologies which run over analogue lines or other means: audiographics, videoconferencing and various forms of remote database access and file exchange.

Audiographics

The term *audiographics* applies to a small range of technologies which combine a live voice link with a shared screen for graphics, real-time drawing or pre-prepared material. It is a technology in transition, partly due to ISDN capabilities, but almost certainly destined to be subsumed by multimedia desktop conferencing. For example, most of the vendors of audiographics systems are adding video windows, integration with other facilities resident on the local machine and, of course, ability to run over ISDN. Another 'add-on' is a projector screen, which enables audiographics to be used in lecture mode to large groups of students. The remote lecturer's slides, graphics or drawings appear on the projector, accompanied by voice and live pointer on the screen. The Co-Learn software, described in Chapter 11 is a clear example of audiographics evolving towards desktop conferencing. Co-Learn combines three forms of audiographic interchange (lecture, small group and one-to-one) with the facilities of an asynchronous messaging system.

In this interim period, audiographics applications are particularly exploited in the primary/secondary sector, and to a lesser extent in tertiary and vocational education. Its cost effectiveness has been proven in a number of studies (Chute *et al*, 1990; Smith, 1992). Its main niche seems to be in extending the course choices of urban students to small, isolated communities. Students who would otherwise have to move to larger centres to receive more specialised courses, can now be accommodated in their own environment. Although certain subject areas like mathematics, technology, engineering etc. would seem to be most appropriate for use of audiographics, in fact there are applications across all curriculum areas. Examples abound in North America, Australia and the UK. Not all of audiographics use is over ISDN, but many of the sites are planning to upgrade soon.

The simplest configuration of the system is for linking two sites – for example, the expert teacher with the remote classroom. More complex arrangements involving an audio bridge can connect several sites – usually not more than four – and hence the possibilities for student-to-student exchanges, peer learning and live discussion can be exploited. The systems are relatively simple to learn and to operate, and most allow any participating site to receive and transmit.

Some would argue that the term audiographics should be reserved for systems which include proprietary hardware and software. However, the facility to share a screen across several sites can now be accomplished without buying in a specific audiographics package, for instance, by adding software which allows one machine to take-over the working of another machine. Combined with facilities already resident on a typical machine (such as a drawing package and word processor) or purchased independently (such as a graphics pad), all of the components of an audiographics system can be put together. One issue which arises with drawing packages rather than a graphics pad, is that it is much easier for new users to work with a graphics pen and tablet than to learn how to manipulate a drawing package. Users who are already familiar with their own drawing package, would find a 'put together' system quite acceptable in terms of user interface. In fact, such a system is another step towards multimedia desktop conferencing.

Whether current systems do, or do not, have a video window (showing the person at the opposite end), there is one main factor which distinguishes audiographics from desktop videoconferencing. Audiographics equipment is typically accessed from a school, a training room or a community centre, where a group of students sit around a table and share the graphics pad. Desktop conferencing is a 'one person per machine' concept, where each machine may be in a different building, city or country. The difference is significant pedagogically, but both have benefits and limitations. Some audiographics courses capitalise on the presence of the group by assigning group tasks, providing 'down-time' during a session for the group to break the connection and work together, or simply by encouraging a social environment for learning. Some teachers take care to identify each student, and call on them by name with specific questions. Interactions amongst the group before and after the audiographics session are also part of the whole educational experience for each person. Despite lack of evaluation evidence from desktop conferencing applications, it is obvious that individuals working at their own machine have the tremendous advantage of control: over their participation, access, input and all the facilities on their machine. Their contact with all other students, however, is mediated by the technology, and any group work, such as shared report writing, consists of individuals working together, rather than groups interacting with other groups.

The limitations of audiographics systems are two-fold. First of all, lesson planning (creating computer visuals) can be considerably time-consuming for the teacher.

> Audiographics slides must be well designed and properly sequenced. Instructors should 'story board' or carefully plan each visual for a lesson before actually creating the slide/visual on the computer. Principles of effective visual communication (balance, harmony, contrasts, lettering, etc.) should be incorporated when creating slides. (Barker and Goodwin, 1992)

Secondly, audio bridge systems are notoriously prone to technical hitches. Losses in transmission connection, extraneous noise or interference on the telephone lines can cause havoc in loudspeaking telephones, and acoustic echo can be difficult to control. ISDN promises to end these technical problems – we await confirmation.

Videoconferencing

Videoconferencing is a confusing term in that it encompasses on the one hand, live video-lecturing to large audiences, and on the other, desktop conferencing point-to-point. The majority of large-scale applications currently are satellite-based – one-way video, two-way audio. Students at a remote site, or sites, see and hear the lecturer on several monitors positioned around the room. Using a telephone, they can call in questions live to the remote lecturer. ISDN promises to make two way video equally as cost-effective but more interactive. With two-way video, every site can hear and see, and speak to, any other site. ISDN also facilitates smaller scale, multi-site videoconferencing, which lends itself to seminars, peer interaction, demonstrations of equipment or procedures, or round-robin discussions and training sessions. Like audiographics, videoconferencing with ISDN shades into desktop multimedia conferencing, and the major suppliers of video equipment all offer room-based systems for large scale applications, 'rollabouts' for medium size flexibility and desktop versions for personalised interaction. One manufacturer even markets its desktop videoconferencing system much like audiographics, with electronic pens in six different colours for marking documents from each site.

If the demarcations between the actual technologies are shifting, the terminology used to describe them creates the ultimate confusion. Almost without exception, manufacturers (and their promotions departments) and to some extent users in different countries choose the same words for different facilities and technologies (e.g. 'teleconferencing' means different things on either side of the Atlantic, 'multimedia' can mean enhanced computer-based training, multiple media, or networked desktop conferencing, and the term, 'virtual classroom' can be based on video lectures or computer conferencing). Equally, different words are frequently used for the same facility (e.g. candid classroom, video lectures, television, teleconferencing, videoconferencing are all used for the same activity).

The diversity of ways of using videoconferencing also contributes to the confusion about what it is and how to use it for education and training. Although referring to satellite-based videoconferencing, a report written by the Audio Visual Centre at University College Dublin, lays out the range of pedagogic scenarios for using video in education most comprehensively (Phelan, 1992). A brief summary will help to clarify the options:

- A lecturer (usually an expert) delivers a lecture from a studio – with graphic or other illustrative material – to students at a remote site or sites. This scenario may die completely with the two-way videoconferencing on ISDN (but see Chapter 9, by Lafon).
- Candid classroom, where some students are present with the lecturer and some are remote. This is the most popular option with the advent of ISDN.
- Mediated presentation – often used in training applications where various experts are used to explain and present material, and a range of artefacts, pre-prepared graphics or complex procedures are demonstrated. Inserted video clips made beforehand or selected from video banks, can be used. This kind of videoconferencing usually requires considerable planning and is closer to what we are familiar with as live television programmes.

Several ISDN training applications of video lecturing and small-scale videoconferencing are described in this book. Chapters 6 and 8 refer to the considerable use of videoconferencing (of the candid classroom sort) in Australia and the US. The benefits are obvious:

- maximisation of lecturing time and expertise
- travel savings for teachers and/or students
- extended course choice for students at smaller institutions
- in some cases, a better view of equipment, procedures, computer graphics and artefacts
- more timely access to information.

Lecturers can do what they have always done, which is lecture (at least in the first two types), and organisations claim it reduces costs. Finally, reports, evaluations and research on applications invariably show that students are positive about learning in this medium. It is no wonder that the literature abounds with predictions that this kind of application will grow and grow (Gunawardena, 1991; Stone, 1990).

If the model one begins with is the face-to-face lecture, then videoconferencing has few disadvantages. With two-way video now more affordable with ISDN, it is almost as good as being there (for spontaneity, feeling of presence and sense of a group), and in some cases, better than being there (for ease of access and a better view of certain procedures or equipment). The limitations are inherent in the nature of lectures and the transmissive model of education. However, with the continuing pressure to increase student numbers without increasing costs, it is unlikely that smaller teacher/student ratios, discussion groups and other encouragements towards deep-level learning will find much financial support. There are indications, however, that one of the most positive outcomes of technology-mediated education and training is that teachers are improving their standard lecturing techniques: preparation of visual aids for audiographics and videoconferencing leads to better lectures and the interactive potential of these technologies has kept lecturers 'on their toes.'

Remote access to information

This third category of applications is admittedly amorphous, but is intended to encompass a range of unique uses of ISDN. Access to data banks of pictures is one example of a facility which can be woven into any number of educational applications. Faster access to textual databases and much speedier file transfer afforded by ISDN mean that a whole range of activities can be used educationally, which were otherwise too slow or too expensive. For example, researchers at museums or hospitals can access and work on banks of image stores at remote sites; similarly, trainees in the workplace can download technical drawings to their workstation. ISDN links are also used in combination with satellite, as the return channel from receive sites, or as the access route to an image server – a bank of images (see Chapter 13 by Beckwith) – during live videoconferences.

Resource-based learning, in which students choose their own path through a range of resources under the guidance of a tutor, is a pedagogical approach ideally suited to ISDN. Institutions can build up a bank of material – such as still images, lecture notes, databases and audio tapes – to make available to students to access in their own time.

> Integrated systems put a plethora of telecomputing
> capabilities at the fingertips of teacher and learner. Both can
> access multiple programs simultaneously, and use different
> application software as they may need. These systems are
> also capable of driving other digital media, such as CD-ROM
> players and videodisc players. There are virtually no limits
> to databases, conceivably worldwide, and multimedia
> libraries of digital sound, text and image. (Saba, 1990)

With careful planning, resource-based learning could resolve the conflicting pressure to increase student numbers while providing personalised, student-led learning and convenient access to education and training. It also has the major advantage of traditional distance education: the learning material can be accessed at times of the students' choosing. Videoconferencing and audiographics courses may be, to some extent, place independent, but they are not time independent. For dedicated distance teaching universities, which have built up a student body used to fitting their studies around a competing range of work, domestic responsibilities and leisure activities, course components based on attendance at a particular place and time are not generally welcomed (Bowser and Shepherd, 1991).

As the market grows, ISDN will undoubtedly be exploited in a variety of novel and unique ways to carry parts of education and training courses. In the meantime, this area of innovation is the most exciting aspect of the advent of ISDN, in that it is being used to provide new educational situations. It is slow to develop because it is not based on standard models, but on a real exploitation of what ISDN uniquely provides to the education and training market.

Learning through interaction

Interaction is not just 'in vogue' in educational circles; it sometimes threatens to become synonymous with the term learning! Without interaction, the teacher cannot know who actually understands the material, and cannot resolve misunderstandings. Without interaction, there is no way of answering unanticipated questions. Without interaction, students cannot express their learning difficulties. These concerns are in fact, only the beginning of the story. Interaction at the question/answer level is a first order use of communication. It does not disrupt the 'transmissive model' of education, which emphasises mastery of a body of pre-structured material. In some learning situations (e.g. skills training), this model may be appropriate. However, much more challenging for teacher and learner is the kind of interactive model of learning in which understanding is achieved through dialogue, commenting on ideas, reformulating ideas, and building on group energy and synergy. The technology which best supports this model of interaction is computer conferencing, because it is asynchronous and users can reflect on their interactions through written communications. Multimedia desktop conferencing may well follow this path, as it provides messaging facilities in addition to real-time interaction. Small group teaching situations, whether technology-mediated or face-to-face, also have potential for real interaction – for active involvement of all participants, for peer commenting and group synergy. Lectures are a different educational activity altogether, and the kind of interaction which typically takes place is question and answer. Interestingly, one researcher reports that students rated feedback on their written assignments more highly than the ability to ask questions in video lectures (Barnard, 1992).

Interaction between student and teacher or amongst students is, in fact, quite a small element in the process of learning. In most sustained education or training courses, whether face-to-face or mediated by technology at a distance, the student will spend more time and put most attention into the isolated activity of 'interacting' with the learning material.

> There is an even greater myth that students in conventional
> institutions are engaged for the greater part of their time in
> meaningful, face-to-face interaction. The fact is that for both
> conventional and distance education students, by far the
> largest part of their studying is done alone, interacting with
> textbooks or other learning media. (Bates, 1991)

Lectures, textbooks, and other printed material are tried and true 'media' for presenting the stuff of knowledge. Just as lecturers have ways of presenting information in a vivid or thought-provoking way, so textbooks can use self-assessment questions, activities and model answers to encourage analysis and integration of new material with existing knowledge. This stimulates a form of interaction in the listener/reader, which is no less valid or important than human to human communication. Some forms of Computer Based Training (CBT) and most Interactive Video Disks are also called interactive by their promoters, meaning that the student can make choices, carry out exercises and procedures to test their understanding and work through material in their own way.

Theories of learning suggest that it is through interaction with ideas and with other people that the student develops understanding. Opportunities for both types of interaction are increasingly considered essential for the best quality provision. Not surprisingly, therefore, most of the applications of ISDN facilitate and support interaction: between teacher and student; amongst students themselves; and with ideas.

Access (via ISDN) to vast stores of text and images is certainly the form of interaction least exploited at the moment by teachers and trainers, although this is probably due to the cost of equipping individual students. Student to student interaction is another under-developed resource which ISDN should be facilitating – audiographics applications are leading the way here. Interaction between teacher and student is by far the most common interpretation of the concept, and it is not coincidental that this model directly reproduces the non-technology-mediated version – the lecture followed by a question and answer session.

A further development of the 'interaction debate' considers that active participation by each student in discussions and group work, should be the norm where technology permits. In the educational use of computer conferencing, for instance, pressure is often put on all students to comment regularly and contribute to exercises and discussions. 'Lurking' is the term used to describe users who just read the inputs of the tutor and other students, and some consider it unacceptable and unfair to the other participants. At the very least, it is thought to be more beneficial educationally to engage in debate, than merely to observe it.

Integrated systems have added a new dimension to the interactivity issue: virtual contiguity. Beyond the kind of interaction that take place face-to-face, there is the interaction around shared resources which need not be synchronous.

> Sharing and multitasking intensify voice and sight dialog
> beyond face-to-face communication. This intensification is a
> significant feature of integrated systems. Its significance does
> not lie in its comparative value of face-to-face
> communication. Its importance is that it enables the teacher
> to respond to the needs of the learner by accessing a variety
> of information sources and making it expeditiously available
> to the learner. (Saba, 1990)

With the multitasking potential of desktop multimedia conferencing, interaction may take an entirely new direction.

Nevertheless, as most teachers, trainers and tutors who have tried to facilitate interaction know well, there is a difference between providing the opportunity for discussion (whether after lectures or in discussion-based courses) and students making use of the opportunity. Equally, there is a difference between the first order use of the opportunity for interaction (factual queries, clarifying misunderstandings and 'clever' questions), and a significant discussion (where ideas are formulated and expressed, and comments on the substance of the ideas are heard and considered). Setting up an educational environment which fosters this kind of interaction is a skilled undertaking, in which the use of any technological media is a minor part of the whole enterprise. Much of what passes for interaction in educational settings may be beneficial, but it does not amount to much more than clarification and information exchange. The educational literature abounds with articles on the new interactive technologies (primarily videoconferencing); however, although interactivity is clearly the major selling point, there is almost never any mention of the nature, quality or even quantity of use made of this facility.

> Interaction does not simply happen with a two-way interactive medium. It is also important for designers and instructors to think of interaction in a broader sense rather than as telephoned-in questions from the sites during class time, and to provide for interactive activities before, after and between class sessions. (Gunawardena, 1993)

Experienced audiographics instructors give the same advice:

> One has to adopt a much stronger approach with students in a telematics link in order to engage them effectively. This 'dialectical encounter' is necessary to keep students involved in worthwhile interaction. It is too simplistic a view to assume that opening a telematics link once a week necessarily engages students in effective dialogue, particularly given the lack of experience of many students with the technology. (Smith *et al*, 1993)

The current vogue for interaction has put non-interactive technologies, such as broadcast TV, radio, video cassettes, and to some extent, CBT, into the developmental shade. Yet, a recent American survey shows that non-interactive technologies are still by far the most common in distance education. What it also shows is the range of student support services used by institutions, which may be as effective as most interactive technologies: telephone office-hours for faculty teaching the course, study guides, individualised feedback from faculty, student access to public or college libraries, pre-broadcast notes and follow-up exercises (Gunawardena, 1991).

It is much too simple to divide courses into those which are interactive and those which are not (and implicitly to regard those which offer human to human interaction as best). When the word interactive is applied not only to video lectures with hundreds of students but also to small groups in an audiographics sessions, as well as to CD-I (compact disk interactive), it is time to find a new word – in fact, at least three words!

It is perhaps also worth referring to the negative aspect of interaction – heretical as that may be in the current climate. Working in groups can lead to group transference, to the 'lowest common denominator' solution to problems, and to time consuming process discussions:

> Collaborative learning is not a pure good. One cost of
> working in groups is that some resources must be invested
> in group coordination. For instance, planning as a group
> takes time and effort, and may have psychological costs.
> When a group member is listening to others, he or she is not
> writing. When a group member is talking to others, he or she
> is not listening. Social psychologists who study small groups
> have called the transaction costs of being in a group 'process
> losses'. (Kiesler, 1992)

If group co-ordination activities take time face-to-face, they invariably take longer when mediated by technology, especially asynchronous messaging media.

ISDN and educational technology

Anyone who has been involved in education or training for very long, has seen many technologies come and go, but each was proclaimed to be revolutionary during its ascendancy. The revolutionary picture conjured up by ISDN enthusiasts involves instructors and students using the ordinary telephone in an office, classroom or workplace, to connect up their desktop workstations. Students can go through an interactive video lesson, work through a multimedia, computer-based package, watch a video segment, take a test and communicate in classes with the instructor and other students about course issues. Dismissing questions of cost, this is all possible now. Aspects of this picture are described in actual practice in the chapters which follow.

Much of the visionary literature about teaching machines, educational television and satellite was also true, and technically feasible. Some of the benefits were even achieved, but the overall prediction of revolutionary changes in the great mass of education and training have not happened. It is unlikely that ISDN will affect education or training on a mass scale. This has more to do with the conservative core of the massive education and training activity world-wide, than it does with educational technologies.

If lack of knowledge about ISDN and how it can be used educationally is a current deterrent to its spread, over-selling of its potential is equally dangerous to its successful spread. This is the main lesson to be learned from the history of previous technologies. ISDN *will* be superseded (Kirvan, 1993; Warwick, 1993); it will not totally transform education and training, but what it will do is integrate voice, text and video onto the students' and teachers' desktop. It does provide a very broad base on which to build education and training applications: large scale videoconferencing, small group co-operative learning, computer-based packages, multimedia interactive sessions, and very fast file exchange and access to learning resources.

Should trainers and educationalists invest in ISDN? The answer is yes, under the following conditions:

- if there is a need for relatively high bandwidth at relatively low cost
- if two institutions want to co-operate on course delivery
- if there are several sites in remote locations all wanting to access courses
- if text-based exchange (computer conferencing) is not adequate for the course content
- if expertise must be shared amongst distant locations.

In short, ISDN offers two facilities: bandwidth and connectivity. For those practitioners who are realistic about innovation, there is another reason to look to ISDN:

- because it is easier to attract financial support for technologies which are at the leading edge.

Why not stick with POTS technologies (Plain Old Telephone Service) and lie in wait for the next educational 'revolution?' This question can be generalised to any technology, and the general answer is that it is harder to maintain a state of 'innovative readiness' than it is to go ahead and innovate. Those who are best placed to capitalise on the next educational technology are not those who have sat back and watched this one go by. What is important to look at with ISDN is the nature of interactivity required – whether surface or deep. Secondly, programmes which are balanced and 'fit for purpose' should be designed such that expensive, high contact (video and face-to-face) sessions are interspersed with reflective interaction (computer conferencing and print). Finally, it must be remembered that the educational objectives of the implementation are the first priority; the technology is but the servant.

References

Barnard, J. (1992). Video-based instruction: issues of effectiveness, interaction, and learner control. *Journal of Educational Technology Systems*, **21**, 1, 45-50.

Bates, A. W. (1991). Interactivity as a criterion for media selection in distance education. *Never Too Far*, **16**, 5-9.

Barker, B., and Goodwin, R. (1992). Audiographics: linking remote classrooms. *The Computing Teacher*, April 1992, 11-15.

Bowser, D. and Shepherd, D. (1991). Student perceptions of the role of technology in enhancing the quality of management education at a distance. In: *Quality in Distance Education: ASPESA Forum* (eds. R. Atkinson, C. McBeath and D. Meacham). Australian and South Pacific External Studies Association, Lismore Heights, NSW.

Chute, A., Balthazar, L. and Poston, C. (1990). Learning from teletraining: what AT&T research says. In: *Contemporary Issues in American Distance Education* (M. Moore, ed.). Pergamon Press.

Gunawardena, C. (1991). Current trends in the use of communications technologies for delivering distance education. *International Journal of Instructional Media*, **13**, 3, 201-213.

Gunawardena, C. (1993). Review of *Videoconferencing and the Adult Learner*. *Open Learning*, **8**, 2, 66-7.

Kiesler, S. (1992). Talking, teaching, and learning in network groups. In: *Collaborative Learning Through Computer Conferencing* (A. R. Kaye, ed.). Springer Verlag, Berlin.

Kirvan, P. (1993). ISDN in the US? *Communications Networks*, January 1993, 17-20.

Phelan, A. (1992). *DBS Pedagogical Scenarios*: The Use of Direct Broadcast Satellite in the Multimedia Teleschool for European Personnel Development. Delta Deliverable 5, Project MTS D2021, Brussels.

Rajasingham, L. (1990). Integrated Services Digital Networks and distance education. *ETTI*, **27**, 3, 301-304.

Romiszowski, A. (1990). Shifting paradigms in education and training: what is the connection with telecommunications? *ETTI*, **27**, 3, 233-237.

Saba, F. (1990). Integrated telecommunications systems and instructional transaction. In: *Contemporary Issues in American Distance Education* (M. Moore, ed.). Pergamon Press.

Smith, K., Fyffe, C. and Lyons, L. (1993). Telematics: the 'language of possibility'. *Open Learning*, 8, 2, 45-49.

Smith, T. (1992). The Evolution of audiographics teleconferencing for continuing engineering education at the University of Wisconsin – Madison. *International Journal for Continuing Engineering Education*, **2**, 2/3/4, pp. 155-160.

Stone, H. (1990). Economic development and technology transfer: implications for video-based distance education. In: *Contemporary Issues in American Distance Education* (ed. M. Moore). Pergamon Press.

Warwick, M. (1993). A technology still in the wilderness? *Communications International*, **20**, 4, 55-57.

Chapter 3

ISDN – strategic issues

Paul Bacsich

Introduction

This paper aims to place the discussion in the last two chapters about Basic Rate ISDN and educational applications in an overall context of educational networks. A subsidiary aim is to give general information about applications of ISDN in the education and training market that have not been included as specific case studies in the book, either because they are not widely deployed as yet or because they affect education in a 'behind the scenes' manner.

ISDN coverage

What countries does ISDN operate in?

ISDN is still not a very pervasive technology in terms of the countries it covers. At the time of writing, there were 19 countries that have deployed an ISDN network, or at least a network sufficiently close to ISDN that it could inter-operate with it. Of these, 12 are in Europe, and of these, eight are in the European Community. Table 3.1 gives the details.

How standard are these ISDN implementations?

Table 3.1 also indicates that in several countries, the ISDN implementations are non-standard to a greater or lesser extent. In addition, some European countries deployed ISDN-type networks on a limited basis to customers, prior to firm CCITT standards being agreed, in order to obtain early feedback from the market. Some of these networks became quite well known, and since the specifications are considerably different from ISDN standards, they still cause confusion, especially in the minds of those (like many educators) less in touch with the technology. In the US, the break-up of the Bell system and the orientation to 56 kbit/s networking has also caused problems.

Country	Network	Basic Rate	Primary Rate
Australia	Microlink/Macrolink	Yes	Yes
Belgium	ISDN	Yes	Yes
Canada	Globe Access 64	Yes	Yes
Denmark	ISDN	Yes	Yes
Finland	ISDN	Yes	due 1993
France	Numeris	Yes	Yes
Germany	ISDN	Yes	Yes
Hong Kong	Dataline	Non-standard	Yes
Italy	ISDN	Yes	Yes
Japan	ISDN INS-Net	Yes Yes	Yes Yes
Netherlands	IDN	Non-standard (standard by 1994)	non-standard
New Zealand	ISDS	Yes	Yes
Norway	ISDN	Yes	Yes
Singapore	ISDN	Yes	due 1993
Spain	ISDN	Yes	Yes
Sweden	IDN 64000	Non-standard	non-standard
Switzerland	Swissnet	Yes	Yes
United Kingdom	ISDN Mercury	Yes No	Yes Yes
United States	Accunet-56 AT&T SDI MCI Bell South and several other Baby Bells	Non-standard No No Non-standard	No Yes Yes No

Note: Some other countries (such as Portugal) have started deployment of national ISDN, but not yet linked to international ISDN.

Table 3.1   *ISDN networks connected to world ISDN*

Europe
Since 1986, the European Commission has encouraged the PTTs to co-operate on a specification of a Euro-ISDN. As well as specifying data rates as in the CCITT Basic Rate standard, the Euro-ISDN standard also covers signalling and a basic tier of services offered. All European countries (to be precise, all those in CEPT) are supposed to change over to this in the next few years.

United States
The ISDN situation in the USA has been very confused with at least a dozen carriers offering services more or less conforming to the ISDN standard, with varying degrees of interworking with each other and international ISDN services. In 1991 agreement was reached on the first draft of a 'National ISDN' standard, and since then there has been good progress towards interworking. In terms of international interworking, this is working for Basic Rate. It is still the case that the USA is much less enthusiastic about ISDN than the Europeans. This is for several reasons:

- the use of 56 kbit/s as the basic building block of digital networks rather than the 64 kbit/s of ISDN (and European networks)
- competition to ISDN from non-ISDN switched digital services
- much greater use of high-speed modems (9600 bit/s and above) than in Europe
- much greater use than in Europe of digital leased circuits at megabit/s rates (T1 – 1.5 Mbit/s) where Basic Rate ISDN cannot compete in speed and Primary Rate ISDN has significant problems of interfacing.

I shall discuss these problems in more detail, and their implications, in a later section.

ISDN applications

You may already have gained the impression that the only applications of ISDN in education and training are videoconferencing and various other forms of teleconferencing. However, as videoconferencing becomes more successful, it is likely to move off ISDN to other networks (such as leased lines). Thus if ISDN is to have a continuing significant role in education, it will have to meet other requirements than just videoconferencing.

I shall begin this section by looking at the initial hypotheses as to the uses of ISDN that were promulgated by the early enthusiasts, especially the PTTs. I shall then move on to look at current thinking on uses of ISDN and current applications. Finally I will draw some conclusions as to the likely uses in the education and training sector, including teaching and non-teaching applications.

Current applications

It is now over eight years since the first ISDN-like service was deployed (the IDA service in the UK). After this period, the original applications ideas of the PTTs have been tested with real customers. Most have not found much favour, some have found favour in revised form, and new ones have appeared as follows:

	Service	*Success*
1	telephony at 64 kbit/s	little interest in telephony
2	audioconferencing	little interest yet in audioconferencing, despite the advantages of ISDN for audio bridging
3	Group 4 facsimile	surprisingly little interest even though the application is an obvious one
4	teletex	service is little used on any network
5	videotex – alphageometric and alphaphotographic	little interest in videotex; increasing potential for access to other kinds of online services
6	computer communication at 64 kbit/s	videotelephony file transmission (of images and other things) linking LANs

Table 3.2   *CCITT teleservices for ISDN – the modern view*

I can summarise the table by saying that the category #6 (which essentially started life as 'none of the above') has been the one where almost all of the current applications lie, with interest (but not much current activity) in #3 and #5. I shall look at these areas in more detail. But first, I shall review the main disadvantages of the PSTN and advantages of ISDN.

The PSTN has many disadvantages. Paraphrasing Lawton (1993), the main ones are shown in Table 3.3 below.

1	PSTN is noisy. This causes loss of quality in phone conversations and errors in data transmission.
2	Call set-up times are long – 5 to 25 seconds. For short data calls (such as for credit card verification), the call set-up time can be longer than the call.
3	The bandwidth is restricted to 3 kHz. This sets an upper limit to the rate at which data can be transmitted, however clever the modem technology used.
4	The characteristics of a call are not fixed, since routing of calls is determined at call set-up time. This makes it hard to obtain optimum performance on a given call.

Table 3.3   *Disadvantages of PSTN*

ISDN overcomes these disadvantages. Thus it would seem that even for telephony, the attraction of ISDN would be significant; but it has not been. Why not? The reason is that the same process of digitisation that has been used for ISDN, has affected much of the PSTN also. Table 3.4 rephrases Table 3.3 in terms of the modern PSTN.

1	Taken over all calls that a user makes, the modern PSTN is perceived as much less noisy. Reasons include: – the trunk network in most advanced countries is all-digital – handset microphones are much better (no more carbon granules) – organisation PBX systems are much less noisy even if non-digital – for data calls, often only one local link (from the originator to a nearby modem) is analogue, because the call is then routed over a data network.
2	The perceived length of time to make a call is much shorter on the modern PSTN. (By 'time to make a call', we mean time from the point at which a user picks up the handset to the time when the called party answers.) There are two main reasons for this: – Using a tone dialling keypad, the average user can dial a 10-digit number in 3 seconds, as opposed to about 10 seconds when using a rotary-dial system. If the number is pre-programmed, this time is even less – about 1 second. – On a modern PSTN exchange (such as System X), call set-up times can be as low as 1 second. These times are now shorter than the time it takes many people to answer the phone. Thus shortening them more is not attractive to users of telephony.
3	Even though there is an upper limit to the speed of data transmission, current modem developments are producing modems with speeds over the PSTN which are almost half the speed of an ISDN connection. And the PSTN is ubiquitous.
4	On the PSTN, clever modems (such as V.fast) will 'probe' the connection to determine its dynamic characteristics and adapt accordingly. On the ISDN, even though the data rate of an ISDN call is fixed, the other characteristics of an ISDN call are still not fixed (such as propagation delays), since routing of calls is determined at call set-up time. This causes problems for developers of systems which require multiple ISDN connections, such as videoconferencing.

Table 3.4   *PSTN disadvantages revisited*

Telephony
Thus as Table 3.4 (rows 1 and 2) shows, for telephony the two main advantages of the ISDN over the PSTN, better quality and shorter call set-up time, are much less significant than commonly realised.

The other claimed advantage of ISDN over PSTN for telephony is that there are various *supplementary services* that ISDN provides. These are similar to those provided on any modern PBX and include:

- call waiting indication
- call diversion
- calling line identification.

However, with the exception of the last, the other services *are* now available on the PSTN whenever a modern digital exchange (such as BT System X) has been installed. And the last service has raised many questions of privacy opt-outs (for example on help lines) which may inhibit its early deployment on ISDN.

My conclusion is thus that, as a service for telephony, the advantages of ISDN are not convincing.

The situation is not dissimilar to that in the domestic market for television – despite many complaints about the quality of the current television standards (PAL etc.), consumers have been very reluctant to trade up to something better.

Audioconferencing
The same worries about telephony are likely to apply to audioconferencing, but to a lesser extent. Audioconferencing is a more specialist service, and the specialists are well aware of the inadequacies of audio bridges over the PSTN. Thus I expect to see some deployment, and some take-up, of ISDN-based audioconferencing systems.

However, audioconferencing suffers from being a 'cinderella technology', overshadowed by videoconferencing. There is massive PTT and vendor investment going into videotelephony for the consumer market, and this is likely to result in cheap videotelephony systems, for both consumer and educational use, before too long. In that era, what will be the role of audioconferencing?

Group 4 facsimile
Facsimile is a technique for transmitting text and black-and-white pictures over the telephone network. The current mass-market standard for facsimile over PSTN is *Group 3*. This operates with 9600 bit/s modems (which drop down to a lower speed on noisy lines). At the resolution used in Group 3, an A4 page is composed of 1145 by 1728 pixels, nearly 2 Mbit of information. For such a volume of data, a speed of 9600 bit/s seems hopelessly slow, but complex data compression techniques are used to reduce the average transmission time of an A4 page to about 30 seconds.

Group 4 is a new CCITT standard for facsimile oriented to use over the ISDN. This offers resolution, grey level, error correction and data compression facilities all better than Group 3. It is designed to allow transmission of a typical A4 page in about 1.5 seconds over one ISDN channel.

Group 4 facsimile was seen by the PTTs as one of the winning applications for ISDN. But needless to say, early Group 4 machines are extremely expensive, and as with telephony, the analogue technology has not been standing still. The perceived quality and speed of Group 3 facsimile systems has been improving over the last few years.

There are also worries that the Group 4 standard is not sufficiently integrated with the corporate world of laser printers and desktop publishing to be attractive in the office context. (Laser printers have standardised on a resolution of 300 pixels per inch; with 600 pixels per inch for more sophisticated ones – these are not the same as the resolution of facsimile systems.) In addition, although facsimile is a very popular technology, in the long term it will be overtaken by electronic transmission of documents, and the dominant formats being proposed for that are not based on facsimile (or on input from PTTs) but from the vendors of desktop publishing systems.

Teletex

Teletex was an enhanced version of telex, based on a concept of communicating word processors, which was proposed by the CCITT. Teletex suffered badly from competition from mainframe-based electronic mail systems, including those offered by PTTs themselves as rivals to teletex! It has never become popular and has been withdrawn from service in several countries such as the UK.

Videotex – alphageometric and alphaphotographic

Videotex is a CCITT term for a standardised teleservice concerned with the transmission of 'pages' of data over telephone lines and its display on modified TV equipment. The service has different names in different countries – Prestel in the UK, Minitel in France, Bildschirmtext in Germany. Videotex has not become popular, with the notable exception of France where Minitel is widespread.

The basic videotex user interface paradigm, as well the technology, is now obsolete (being overtaken by PC-based windowing approaches and client-server systems respectively); so that in my view it is too late for ISDN to rescue videotex in its current form. Nor will videotex be the winning application for ISDN.

However, various screen-based services are likely to be deployed over ISDN, which could be regarded as distant cousins of videotex.

Computer communication at 64 kbit/s
The CCITT saw this as including transmission of images, picture mail, videotelephony and audiotex (online access to a database of audio messages). A modern rewording of this would be as follows:

- videotelephony
- file transmission (of images and other things)
- linking LANs.

Videotelephony (including videoconferencing) is obviously one of the success stories of ISDN, as many of the articles in this book demonstrate.
    File transmission and linking LANs are treated in the next section.

ISDN – a modern view of non-video applications

I begin with linking LANs because this has been one of the largest applications areas yet one of the least foreseen. (As said earlier, the growth of LANs was not foreseen by the early ISDN planners.)

Linking LANs

Many organisations require their Local Area Networks at the various organisational sites to be permanently linked together into a Wide Area Network to support computer applications such as electronic mail and database access.The normal way of doing this (currently) is to use a digital leased circuit, usually (in Europe) at 64 kbit/s. Once LANs are linked together, many organisations rapidly come to depend crucially on the links – they become 'mission-critical'. Thus any break in the link is a serious problem.

Leased line back-up

Many vendors now offer an ISDN Leased Line Backup Unit. These work as follows.
    On the LAN at each end of a leased line link, a Backup Unit is deployed, which continuously monitors the state of the leased line link. If the leased line link becomes faulty, one Backup Unit detects this and makes an ISDN call to the other, which re-establishes connection between the LANs, but now using ISDN rather than the leased line link. The Backup Units continue to monitor the leased line link and as soon as it becomes usable again, the ISDN call is dropped and LAN traffic reverts to using the leased line link.

Leased line top-up

A similar principle can be used to provide extra capacity when a leased line becomes congested. In many organisations, there are times of the day, or days in the year, when WAN traffic reaches high levels. Examples could be:

- first thing every morning when everyone logs in and reads the electronic messages that have come in overnight from overseas offices
- the last two hours on Friday afternoon before a holiday weekend when everyone is trying to clear their desk (by pushing the work on to somebody else's)
- the week before a controversial shareholders meeting.

An ISDN Leased Line Top-up Unit is in its essentials the same as the Backup Unit described above, except that it cuts in and makes the ISDN call when the queue of data waiting to go across the leased line link gets above a certain level, and drops the call when the queue decreases. However, the principle is easier than the practice – it requires quite subtle work to ensure that the Top-Up unit:

- makes the ISDN call when and only when it really needs to
- drops the ISDN call when traffic has 'genuinely' decreased
- avoids making lots of short calls one after another.

Leased line replacement

The most interesting area concerned with ISDN and leased lines is not the back-up or top-up of leased lines, but their *replacement*. Why have a leased line at all when ISDN can offer the same facilities?

Even though ISDN and a digital leased line can offer the same data rates, there are marked differences in tariffing. A leased line is charged on a fixed annual basis – however much or little it is used, the cost is the same. An ISDN line does have an annual rental, but most of the cost of the ISDN service arises from call charges – the more it is used, the more is charged. Charging is a very complex subject, but as a rough rule of thumb in the UK, the annual cost of a 64 kbit/s digital leased line corresponds to about two hours per working day of ISDN usage (one B-channel). Thus the nub of the problem of replacing leased lines with ISDN comes down to the following:

Where are the business applications that require only two hours per day of network connection?

This is not an easy question to answer.

One application that has been suggested is use by stores to send the day's sales figure to head office. This is reasonable enough, but one has to ask:

a Is this the right technological approach (technocrats would argue that it sounds more like an electronic mail or file transfer application)?
b Would the store not need to be online to head office in the day as well?

If one aggregates the two hours per day together, one gets 10 hours per week or 500 hours (about a month of working days) per year. Occasional-use situations that come to mind include:

- a branch office of a weekly newspaper or monthly magazine
- use at a conference or exhibition
- students at an evening class in a study centre (but two hours per day is less than the time that a study centre would be open in the evening).

None of these sound like the basis for a large market. The last is the most interesting to the educational sector. Since the two hours per day rule of thumb is very rough and depends both on the country in question and the distances involved, I feel that it is the one worth exploring in this context.

Thin-route networks

Another area that is being currently explored is the use in organisations where the WAN links are very lightly loaded, that is, do not transmit very much data in the course of a day. Two hours of connection per day would allow 10 minutes data transmission each hour between 8 a.m. and 8 p.m. In 10 minutes (on a B-channel) one can transfer 4.8 Mb, about four floppy discs of information. This corresponds to a large number of messages and reports every hour. (As general guidance, the typical electronic mail message is less than 1 kb in size; the typical memo less than 10 kb; the typical short report less than 50 kb. Few users generate more than 20 messages per day.) Thus many small offices (say with less than 10 staff) could be happy with this grade of link.

The problem is that current methods of linking LANs require that equipment[1] on each LAN is in frequent contact with equipment on all the linked LANs. (For example, on a LAN with Macintosh equipment on it, a certain packet of information has to be sent across the link every 10 seconds.) This has the effect that when the ISDN call is made, it has to be maintained *permanently*, all the time that the LANs are linked – if the call is dropped, a few seconds (or at best minutes) later the LANs will report that services have been discontinued because the LANs are no longer linked.

---

[1]Note to the technocrats: the main culprits are routers.

In order to circumvent this problem, vendors are giving attention to 'spoofing' the LANs, in other words, using equipment that will pretend to the LANs that they are linked when in fact they are not. If they absolutely have to be linked (such as for file transfer), at an appropriate time the spoofing equipment will make an actual call, keeping it as short as possible. (For example, mail and file transfers will be queued up, waiting for the next ISDN call.) All of this is much easier said than done; but progress is being made. If successful, this could open up a large market for ISDN.

High schools and colleges, and other parts of the education sector with modest networking needs could be interested in such products. (On the whole, only the universities are well served with leased lines.)

Modem replacement

The other main non-video area of application of ISDN is likely to be the replacement of modems in those applications where modems are currently used but where modem speeds are not very adequate. Two areas can be distinguished:

1 access to online systems
2 remote user access to LANs.

Regarding the first of these, a common use of modems today is for users to access various kinds of online systems. The commonest types of these are:

a mainframe-based electronic mail systems
b computer conferencing systems
c online databases
d modern hybrid bulletin board/database systems with windowing user interfaces.

Electronic mail systems based on large central computers are run by many organisations: PTTs, Value Added Networks, companies and universities. As an example, the Open University runs a centralised electronic mail system on a cluster of Dec Vax computers, and several hundred staff use this from out-of-office locations (home or travelling). Traditionally, access to such systems has been via 2400 bit/s modems (or even slower in Europe), although access at 9600 bit/s is beginning to become popular especially in the USA. These electronic mail systems tend to have a very traditional user interface – a command-line interface where interaction is organised by typing a series of commands – and messages tend to be relatively short. There is no advantage in using modems with speeds greater than 9600 bit/s on such command-line interfaces because this is already much greater than even the fastest reading speed. In menu-based systems such as videotex, at 9600 bit/s a 24 x 40 character screen fills in one second! This is much less than the average thinking time and further speed increases would be foolish.

Similar considerations apply to type (b), computer conferencing systems. Again, the user interfaces tend to be command-line or menu-based and the information text-based. (Even in the world of desktop publishing, electronic messages are created quickly, being the equivalent of telephone messages more than memos or polished reports, and it is debatable whether they would contain much graphics even if the tools to create graphics were more available.)

Database systems are now moving rapidly towards containing multimedia elements, and so it might be thought that here was a classic application of ISDN. However, most online databases are moving rapidly towards use of CD-ROM, so that there now has to be a strong reason for an online database still to *be* an online database. The most common reason is that the material is changing rapidly (usually daily or weekly) and even in today's world, that tends to imply heavy use of text (news wire services, seat availability, etc.) and little use of anything else.

Type (d) systems offer the most interesting potential for use of ISDN. A typical example is AppleLink, which has the following components:

- an electronic mail system (with, of course, links to facsimile)
- a series of bulletin boards for many types of Macintosh user: software developer, network specialist, etc.
- databases, both of information on Macintosh products (and this changes rapidly – almost every day there are several announcements of new products or upgrades) *and actual software* (bug fixes, shareware, demonstration versions, etc.).

Such systems have two main differences from the command-line systems discussed previously:

1 Navigation involves browsing through windows[1] containing lists of items.
2 Using the system involves downloading information.

A typical large list in such a system may have 200 entries. Until it is transmitted in total, the user cannot rationally make the next choice. If each entry in the list requires 50 bytes of information (for the text and

---

[1]It is commonly believed by educators that if an on-line interface has windows and icons in it, then the windows and icons must have been transmitted *as graphics* from the central system. This is then used to justify the need for high transmission rates. Although there are some systems where this is in fact done (X-Windows systems), the commonest approach is to transmit *instructions* to create windows and icons – such as 'Display icon 234' – but not the windows and icons themselves. These instructions take only a little more time to transmit than text. And all such instructions, text and formatting is transmitted in compressed form anyway - thus in the above example the word 'Display' will not be transmitted but some code such as '$D'.

formatting), this is 100 kbit of data. At 2400 bit/s this would take 41 seconds; at 9600 it would take 11 seconds. Only at 64 kbit/s transmission is the transmission time acceptable at under two seconds.[1]

Downloading is the name given to copying a file from a central system to a user's microcomputer. Software experts, academics and hobbyists are getting very used to needing to download information, including computer programs. Ten years ago, many microcomputer programs interesting to users were under 50 kb in size. At 2400 bit/s, such a program would take under four minutes, an acceptable delay. Now, the programs that users want to download are at least 10 times that size – a Macintosh program such as Excel will now be over 1 Mb in size. Data compression techniques could be used, and are, to reduce the size of such programs, but are not always used and in any case the compressed program may well still be over 250 kb in size. In the last 10 years interesting programs have grown about 10 times in size, users are more impatient, and online charges per minute have risen rapidly. The conclusion is clear: online data rates need to increase. Modems are unlikely to provide the answer: they may do today, but programs are still increasing in size.

My conclusion is that such windowing bulletin board/database systems are a growth area and a good market for deployment of ISDN technology.

Remote user access to LANs

This is the second main potential area for modem replacement. Some explanation is needed first.

In the last few years the use of LANs has become prevalent in organisations. Initially LANs were a way of connecting users to central services and left the services essentially unchanged, but the march of LANs has led and is continuing to lead to a change in the type of system that users handle. The trend is towards client-server systems – in these the balance of work is shared out more equally between the user's microcomputer and the central service – indeed, the central service may run on a microcomputer similar to and little more powerful than the user's, and there may be many 'central service' machines, all interworking.

This puts the remote user of central services in a quandary. There is no longer a mainframe running a command-line interface with which the user can interact using a modem. What to do?

---

[1]In case you are wondering why these figures do not correspond to the transmission figures I gave earlier for videotex screens, the reasons are that (a) modern windowing screens contain far more information than videotex screens and (b) the lists are longer because no one has time to create the hierarchical menu choices, no more than 7 or 8 at a time, that conventional user interface psychology tells us is 'necessary'.

The answer is to 'export' the LAN to the remote user. Vendors have produced simplified versions of the LAN protocols so that a user, at home or travelling, can use a modem to link to the LAN and have a full range of services available 'just as if they were in the office'.

The best known and most effective version of this has been developed by Apple for use on Macintosh networks. AppleTalk Remote Access (ARA) allows users with a modem to access the full range of AppleTalk service on a Macintosh network. But modems have limitations. The following anecdotal information gives a rough guide to performance.

- At 2400 bit/s, AppleTalk Remote Access works, technically, but most actions are too slow to be safely usable.
- At 9600 bit/s, it is usable to do real work. Electronic mail (Microsoft Mail) is fast enough to use, small files can be transferred to or from file servers, and most administrative actions (such as selecting a different printer) can be handled. It is not sensible to print across the link (PostScript files are notoriously large) or to open an application on the far side of the link.
- Using an ISDN Terminal Adaptor at 38.4 kbit/s, users report it as "nearly as good as LocalTalk". (LocalTalk is the cheap networking system available on all Macs – it has a rated speed of about 250 kbit/s but this is shared between all machines on the network segment.)

Note that AppleTalk Remote Access contains its own data compression scheme (with at least 2:1 compression on text) and that the protocol has been simplified wherever possible from the version normally used on LANs.

Similar developments have been undertaken for LAN software that is used on PC networks and are under way for the TCP/IP protocol.

This argument can be summarised as follows:

- In the next few years, more and more users will want to access their host LANs from home or while travelling. (This will be especially true for academic staff.)
- The systems that used to permit remote modem access to such staff are being withdrawn and replaced by client-server systems that require LAN access.
- Thus staff will want LAN access via modems.
- But modems are not fast enough. (When one recalls that the typical LAN operates at 10 million bit/s and even a fast modem at about 10 thousand bit/s, it is a wonder that anything works over a modem!)
- Thus ISDN has a role to play. (A speed of 64 kbit/s seems to be fast enough.)

Home access

This last application brings us to the tricky topic of ISDN services to the home. Much has been written on this topic. Table 3.5 gives examples of online services potentially attractive to home users:

Service	Description
Home catalogue shopping	Online database, with pictures, of goods that can be purchased at home.
Home travel	Online database, with pictures, of holidays, plane tickets, etc. that can be booked.
Home house buying	Online database, with pictures, of houses that can be purchased.
Home banking	Ability to conduct financial transactions (paying bills, etc.) at home.
Home tele-education	Online database (including computer conferencing system) of materials that can be studied at home.

Table 3.5 *Possible ISDN services for home users*

It would take a separate article, and is outside the scope of this volume, to look at the market research in this area, analyse the trials and understand the reasons why this much-predicted home market has by and large not arrived – neither with ISDN nor with regular modems.

Reasons include the following (each with a brief example or two):

- Many of the proposals solve only part of the problem:
  - home banking does not deliver currency
  - booking a holiday is a very small part of the overall holiday purchase activity
- They fail to take into account societal factors:
  - shopping as a recreational activity
  - increasing need for human contact in a mechanistic society
- They go against established business groupings:
  - travel agents
- Some applications have been dealt with by other technologies:
  - automatic teller machines
  - a vast growth in shop by phone where the sales assistants have massive automated support
- Others probably do not need ISDN in order to succeed:
  - a database of flight availability does not need pictures of planes
- Some of them are targeted at groups (elderly, less well off) least able, financially or otherwise to cope with the service.

Home use or not?

Putting all this together, my predictions on home use of ISDN are as follows:

- Large-scale ISDN use in the home will not come from the 'classical' proposals (home shopping, etc.).
- A useful bridgehead in the home will come from professionals working at home on microcomputers compatible with their office systems who wish adequate connection with the office LAN. They are most likely to use Terminal Adaptors rather than plug-in ISDN boards because it is much easier to install and configure a Terminal Adaptor.
- Larger-scale longer-term use will come from the use of videotelephones in wealthier and/or more business-oriented households.
- Group 4 facsimile will not be relevant to home use (even if it eventually gains acceptance in small offices).
- Enhanced telephony services will not be interesting for home users.

None of this, at first sight, makes particularly attractive reading for devotees of home tele-education. However, given careful targeting of courses (MBAs, information technology, etc.) one should be able to establish an ISDN-based tele-education bridgehead in the home in the next 5 years.

Implications for educational use

Gathering up the information in the previous parts of this section, what does this say about prospective educational use?

- Videoconferencing and videotelephony should flourish over ISDN. Education will be able to take advantage of the technology as its capital cost reduces. This niche will include the wealthiest home-based users, who, in Europe at least, are not those traditionally oriented to distance education.
- ISDN access to specialised professional databases/bulletin board systems will be a niche market, but an important niche. Education should be able to tap into this. This niche will include some home-based users.
- Smaller education sites, including schools, will be able to use ISDN to connect to larger networks, at universities, and thus gain access to the Internet and all its services. I see this as an important growth area in several European countries over the next few years. (The USA will achieve the same result but not necessarily with the same technologies.)

This implies that the following ISDN-based technologies relevant to education should benefit from the rapid cost-reduction brought about by economies of scale:

- Terminal Adaptors to replace modems
- ISDN units for occasional LAN linking
- videotelephones and videoconferencing systems.

The following technologies do not look like benefiting from rapidly reducing prices and should be avoided unless a very strong case (including financial factors) can be made for their use:

- ISDN voice telephones
- ISDN-based audio bridges
- Group 4 fax systems.

ISDN enhancements

In this section I re-visit the issue that concerned Jacobs at several points in Chapter 1:

> "We know that ISDN is fast, or at least faster; but is it fast enough?"

To which the answer has to be "fast enough for what?". I shall leave video on one side, since there is much evidence that for certain applications, more than 128 kbit/s is needed, and rather concentrate on aspects to do with discs, printed pages and colour images.

Disc sizes range from the 1.44 Mb (11.52 Mbit) on a dual-density floppy, up through the 5, 20, 40, 80 and 120 common on user micros. (In my world, 120 Mb on one's micro is still a lot.)

Screen image sizes have been discussed by Jacobs (see Table 1.2). Doing the sums, a 640 x 480 x 8-plane image (typical of many computer displays) requires 2.46 Mbit of storage; while a 1280 x 1024 x 24-plane high-resolution naturalistic colour image requires 31.46 Mbit.

Output on paper is still very important in the education and training sector. Much output is at 300 dpi – dots per inch. On an A4 page (size of 297 x 210 mm – 11.69 x 8.37 inches), a 300 dpi image takes up 8.81 Mbit. However, even desktop published output is moving up to 600 dpi. At such a resolution, an A4 page takes up 35.24 Mbit.

Traditional printing has been happy with output resolutions of 2400 dpi for many years, and certainly the human eye cannot perceive differences of that fineness. At 2400 dpi, an A4 page takes up 563.84 Mbit. I regard that as a natural limit.[1]

---

[1]In the analysis of technology development, we come up against two camps. One group says that the natural limits (caused by physical and/or physiological factors) will be important upper bounds on network needs; the other camp says that human wishes will always cause increased demands. We can apply these two points of view to output on paper. One can always ask for larger sheets of paper.

What does all this mean in terms of ISDN transfer times? Table 3.6 lists the answers in terms of four transfer rates: Basic Rate (0.064 Mbit/s), switched-384 (0.384 Mbit/s), Primary Rate (2 Mbit/s) and a proposed entry level B-ISDN rate (34 Mbit/s).

Object – description	Size – Mbit	Transfer time – seconds			
		.064	*.384*	*2*	*34*
Screen medium-quality (640 x 480 x 8)	2.46	38	6	1	0.1
Page A4 300 dpi	8.81	138	23	4	0.3
Floppy disc	11.52	180	30	6	0.3
Screen high-quality (1280 x 1024 x 24)	31.46	492	82	16	0.9
Page A4 600 dpi	35.24	551	92	18	1.0
Small hard disc (20 Mb)	160	2500	417	80	4.7
Page A4 2400 dpi	563.84	8810	1468	282	16.6
Large hard disc (120 Mb)	960	15000	2500	480	28.2

Table 3.6   Transfer times over ISDN Basic Rate and higher-rate services

It is clear from the above table that for Basic Rate ISDN, the time taken for various actions is considerably longer than users currently expect. For example, consider backing up a floppy disc across a network, or loading it from the network. A time of three minutes (180 seconds) for that would seem long. Yet nowadays such a floppy disc would have room for little more than the text and diagrams for this book. As another example, a modern desktop laser printer will print up to eight pages per minute. If the pages are taking over two minutes (138 seconds) to download from the network, that seems the wrong way round – laser printers are supposed to be the bottleneck in the system.

For the screens one may argue that data compression can be used. However, if any losses are introduced then this may reduce the ability to process the information locally, for example by zooming in, cropping and rotating. Even if a 10:1 compression ratio is used, the transmission times are still long – nearly four seconds for a medium-quality screen and 49 seconds for a high-quality screen – on systems where normal user interface design would expect a response in a second or so. In addition, there are audiences, such as doctors, who will be hard to convince that vital information has not been lost in transmission if non-reversible (that is, lossy) data compression is used. (Printers would be similarly sceptical if lossy data transmission was to be used on their precious printing plate images.)

Primary Rate

The natural conclusion is to trade up to Primary Rate. At 2 Mbit/s, a medium-quality screen is transmitted in one second (fast enough for a truly interactive service) and a floppy disc can be copied in six seconds (faster than the information can be read from the floppy). Even an uncompressed 600 dpi A4 page is received in 18 seconds, in other words about three per minute, which is quite close to the actual speed at which laser printers can output different pages.

However, Primary Rate has many problems associated with it. The following can be only a brief introduction to a complex subject.

- A Primary Rate interface is not one channel but a collection of 30 B-channels (plus a D-channel). In order to use it as one channel, complex and expensive inverse multiplexing techniques are needed. This does not make it sound like an approach for early deployment in the home or even in education.
- At current tariffing rates, Primary Rate costs per minute are 30 times the cost of a phone call. Thus in the UK even a local call at daytime Primary Rate would cost 30 x £1.89 = £56.70 per hour! (Plus VAT, of course.) This is hardly likely to endear it to students, academics or indeed professionals working at home.
- The technology used for Primary Rate is much more expensive than that used for Basic Rate.

In other words, pending some startling technological breakthrough or a tariffing breakthrough (less likely), use of Primary Rate does not seem relevant to most of the actual ISDN applications (as opposed perhaps to the fanciful ones).

H0 channel

It has never been very clear why there was such a big jump in network offerings from 64 kbit/s up to 2 Mbit/s. Certainly it is traditional in computing to move up by factors of 10 if possible, but this is a jump of over 30, and in any case, modem speeds move up by factors much less than 10. I suspect that tradition (rationalised as engineering reasons by the engineers in charge) has played a large part. However, there are some interesting plants growing in the 'network desert' between 64 kbit/s and 2 Mbit/s.

The Americans entered the desert the earliest, of course, with so-called 'fractional T1' offerings. These are digital leased lines with a speed between 64 kbit/s and 1.5 Mbit/s (T1 speed). Fractional T1 was oriented to those corporate users who wanted more bandwidth than 64 kbit/s but could not afford trading up to a 1.5 Mbit/s link. Unlike technological development, user developments, such as bandwidth needed, move up by percentages, not orders of magnitude.

To nobody's surprise, fractional T1 has proved very popular in the US. Somewhat belatedly, similar offerings, called fractional E1 by the technocrats, are beginning to appear in Europe.

The speed of 384 kbit/s is often used for fractional circuits. It is a good compromise between 64 kbit/s and 2 Mbit/s (or 1.5), being 6 times faster than 64, with 2 Mbit/s being about the same faster again. In other words, a kind of splitting the difference. Videoconferencing users have also found 384 kbit/s a good speed for videoconferencing systems.

In addition to leased circuits, a number of PTTs in Europe, led by Mercury, have introduced a switched fractional service. This is called Switchband and again operates at 384 kbit/s. The target applications are videoconferencing and high-speed file transfer.

Minoli (1993) reports that there is interest in American ISDN standards circles in so-called H channels. There are several varieties proposed, of which the basic offering is the H0 channel at 384 kbit/s. He also reports interest in a new service offering to be called H + D.

In my view, an H0 + D service offering would be extremely attractive to many organisations, for support of teleworkers and small remote sites. If necessary, to avoid embarrassing comparisons with telephone tariffs, I would be happy if it did not carry telephony at all; but possibly an H0 + B + D offering (which would allow one ISDN phone line) would be more soothing to those PTT elements who believe that ISDN is crucially related to telephony.

Broadband ISDN

Even though it has got off to a shaky start in the market, the ISDN concept has a lot of life in it from the engineering standpoint. Massive development work is under way to produce a *broadband* version of ISDN.

The term "broadband" is hard to define. Many people regard the boundary between broadband and narrowband as occurring at 2 Mbit/s; however, 2 Mbit/s is now not regarded as fast, even for wide area networks. A more modern approach is to regard the boundary as occurring somewhere around 20 Mbit/s – under this classification all commonly occurring local area networks (Ethernet and Token Ring) are narrowband; only the newer LAN standards such as FDDI (at 100 Mbit/s) would be broadband.

Broadband ISDN must be distinguished from Primary Rate ISDN which works at 2 Mbit/s but is 'only' a bundle of 64 kbit/s ISDN circuits, not a fully integrated network. Broadband ISDN is an integrated, bit-rate-independent network which offers service at any bit rate the user requires.

The two main proposed broadband ISDN rates are 150 Mbit/s and 600 Mbit/s. This is far higher than current ISDN rates and far higher than can currently be carried over copper wire networks. Realists are pushing for an entry-level broadband rate at 34 Mbit/s.

Implications for home users (and education)

I have argued that Basic Rate ISDN is not fast enough. Primary Rate ISDN is not really an integrated service and is far too expensive. What about broadband ISDN?

Many people argue that broadband ISDN is very relevant to the home. They point to all the video applications – video libraries, many channels, interactive video databases. Once video is enabled, the data services can piggy-back on it. "All that one needs", they say, "is to put fibre into the home."

Before accepting this argument it is necessary to look at the progress of the precursor of fibre-based video distribution, namely cable TV.

Many European countries have well-developed cable TV systems. In some countries like Belgium the vast majority of the population are connected to cable TV systems. There are more than 29 million cable TV homes in Europe (excluding ex-USSR), more than double the number of satellite TV homes.

However, the reality is that many countries such as France and the UK have only a small percentage of their homes connected to cable – in France just over one million homes, in the UK under half a million. In addition, even in countries with very high coverage, many rural areas do not have cable. Although cable installations are growing, there is increasing competition from satellite TV systems, especially in the UK and Germany.

Possibly if the PTTs went into co-operation with the cable TV companies, a massive fibre installation programme could be undertaken. But why? If a home has cable, they are receiving video already – they will not see the point of having fibre (and paying for it) in order to do what they can do already. The arguments about interactive services will seem very subtle and are not likely to convince them. (The argument is about as weak as voice telephony being the reason to have ISDN.) If a home does not have cable, it may have satellite, or it may not be interested in additional sources of video. (There is also the video hire shop nearby, perhaps.) Either way, it is not a fruitful home for fibre arguments.

Thus I do not think that the arguments for installing fibre are going to be very convincing to the customers.

In addition to the consumer arguments, the idea that the cable companies and the PTTs are going to co-operate is not very convincing. Already in some countries such as UK and the Netherlands there is great concern in the cable industry about the plans of the PTTs. There are also considerations of industrial policy (de-regulation, monopolies, etc.) that are relevant.

My own view is that the best policy is to develop a 'stretched' Basic Rate ISDN technology. After all, the technologies used in Basic Rate ISDN to transmit data over telephone cable go back over a decade. In the last few years there have been considerable advances in transmitting very high speed data over twisted pair cable – not just at Ethernet speeds (10 Mbit/s), which is now routine, but all the way up to 100 Mbit/s. It is known that BT are interested in transmitting broadcast-quality video down telephone cabling and some experiments have been carried out.

The stretched ISDN would have characteristics something like the following:

1 ability to use one pair of telephone cabling from home to local exchange
2 data rates of at least 384 kbit/s in both directions: ideally a little more so that an H0 + B + D interface can be provided (this would be an aggregate of 512 kbit/s assuming a 64 kbit/s D-channel)
3 data rate of at least 1 Mbit/s in the 'downstream' direction so that one quasi-broadcast-quality TV channel could be transmitted as well as the data circuits.

It would not be a complete tragedy if the service required four wires (two twisted pairs). In several countries an increasing number of higher-status homes have more than one phone line already and PTTs are used to supplying spare twisted pair capacity when they install new cabling.

In conclusion I should stress that this issue is not just a European one – similar arguments are being heard in the US about the best way to provide interactive services to homes on a mass scale.

Acknowledgements

Much of the work drawn on for this chapter has been funded by DG XIII of the Commission of the European Community under the following projects:

JANUS is a research and development project under the DELTA programme which has deployed a pre-operational satellite-based network service for distance education and training in Europe, operating at ISDN speeds and integrated with terrestrial networks including academic networks, PSTN and ISDN. The market research phase of this project in 1992 necessitated carrying out an extensive study of the various network options available to European educators.

CCAM was a series of four interlinked study projects funded by DG XIII which studied the potential of telematic networks in various sub-sectors of the education and training sector. Various different network options were costed.

Versions of this paper have been delivered at seminars organised by the European Association of Distance Teaching Universities, the Telecommunications Managers Association and the UK National Council for Educational Technology. I am grateful to these organisations for their support and useful feedback.

References

Lawton, L. S. (1993). *Integrated Digital Networks*. Sigma Press, England.
Minoli, D. (1993). *Enterprise Networking: Fractional T1 to SONET, Frame Relay to BISDN*. Artech House, Boston.

VIDEOCONFERENCING APPLICATIONS

The collection of material in this section on videoconferencing represents something of the range of uses and pedagogical approaches possible with this new medium.

Chapter 4 by Borbely describes Columbia University's Video Network to serve the needs of engineering professionals. He includes details of induction methods for lecturers using the system, as well as thumbnail sketches of three common types of lecturing approaches. The last section, entitled Outlook, gives a sober but challenging view of the future of electronic distance education.

The location for Chapter 5 is Ulster in Northern Ireland. The videoconferencing equipment described by Purcell and Parr is British Telecom's VC7000 and in this application, it is used for remote students to take a course from the University of London (a joint course given by University College and Imperial College). These two authors conclude that digital videoconferencing offers an array of benefits over its analogue counterpart, which are not being effectively utilised. We are still using video to mimic the classroom, rather than building on its unique characteristics. The fact that the medium demands so little change, is exactly why it is so successful!

Chapter 6 by Latchem *et al* gives an overview of the use of ISDN videoconferencing in Australian higher education. Australia clearly has the most developed use of educational ISDN anywhere in the world to date. For Australians the problem is not connectivity but inter-connectivity! The chapter contains a comprehensive listing of codec systems currently in use, as well as a summary of advantages and disadvantages of videoconferencing from the teacher's point of view.

Norway is the location of the next application of videoconferencing, except that the Norwegians call it videotelephony. Perhaps because of this view of the technology, several innovative uses of the technology have been developed in the education/training sector – for example, teaching sign language to the deaf, just-in-time training for groups in the workplace and remote supervision of trainee nurses. Kristiansen illustrates the equipment specially designed for educational use and it is interesting to compare its specifications, both technically and pedagogically, with the French equipment described in Chapter 9.

Chapter 8 looks at the cost benefits of videoconferencing for company training and university distance education, and particularly for partnerships between the two. The range of equipment available is described and a very detailed specification list of videoconferencing features is given with advice and tips on recommended features. Finally, Lange provides a detailed cost benefit analysis for both company and university purchase of a videoconferencing system.

Lafon, in Chapter 9, uses the term 'visiocentre' for the system developed in France, and the equipment and design for its use are quite distinct from all the other systems described so far. In fact, in some respects, the design of the French system is quite the opposite of what other systems seek to achieve. Lafon describes the 'presidential' approach to lecturing and control of interaction, which involves different technology from systems which aim for open access in speaking and viewing (see for example, the description of LIVE-NET in Chapter 13). Lafon's system provides every facility for the lecturer to make a televisual presentation from the concentrated, quiet atmosphere of the Professor's office.

Those readers attuned to the student perspective on videoconferencing will find it interesting to compare the reports in each of these chapters on student feedback and attitudes towards the medium. Comparability of final exam results often show that students do just as well with videoconferencing (or computer conferencing or audiographics) as face-to-face taught students, and many evaluators use these course outcomes to promote the various interactive technologies. However, Borbely in Chapter 4, reports the same comparability for students taught at a distance with videotapes of lectures sent through the post. Is this a justification for extending non-interactive technologies, such as correspondence courses, broadcast programs or videotapes? Quality of educational experience is surely a more relevant criterion, and we should look for evidence of this in student feedback. In fact, Latchem *et al* claim that videoconferencing is not equally effective over different teaching and learning situations, and that the critical factor is the quality of presentation and communication by the lecturer.

Chapter 4

Challenges and opportunities in extending the classroom and the campus via digital compressed video

Edward Borbely

Introduction

Despite the information-rich environment promised by ISDN, there has not been the ubiquitous deployment in the United States that many had hoped for during the 1970s, when most of the required technology was developed. Nevertheless, digital services are widely available today in the USA where the market has justified investments in equipment by local exchange carriers. Digital telecommunications lines are widely available in suburban environments as well; still, the average residential telephone customer has no idea what ISDN is or what it might do. Now that American telephone and cable television companies are showing signs of co-operation, it seems more likely that there will be a jump directly to Broadband ISDN (B-ISDN) for residential customers, since the economics of new telecommunications services in the US residential market appear to be driven largely by entertainment services which require higher data rates.

The level to which ISDN has been deployed, however, hides an important fact: most of the applications we associate with the Integrated Services Digital Network are already widely available, many of these using digital network access. Digital lines of various types have been available in every major US city and in most suburban settings for several years or more, although ISDN per se has seen only marginal success.

Witness Columbia University's graduate-level distance learning programme. In 1986, digital links with AT&T Bell Laboratories and IBM allowed students at these companies to participate in Columbia courses via two-way video and audio. The technology available at that time called for expensive, dedicated, point-to-point connections, highly dependent upon a cast of technical support people in the classrooms and in telephone network control centres. Although interest at these companies was more than sufficient to sustain the programme, not many corporations were willing or able to duplicate the investment.

Technology has advanced to the point where a programme requiring large amounts of money and influence to launch in 1986 can now be implemented independently and at a fraction of the cost. The improvements have been realised both in terminal equipment and in the telecommunications network itself. The presence of digital networking and inexpensive, low bit-rate videoconferencing equipment has empowered individual academic departments and mid-level industrial managers to join forces without the need for major capital investment or central management involvement. It is this sort of empowerment that ISDN has come to stand for, even if (at least in the USA) the precise protocols and standards implied by the acronym have not yet become the basis of our entire telecommunications network.

This chapter will cover the ongoing evolution of Columbia University's Off-Campus Graduate Education Program and the Columbia Video Network (CVN) from the original base of assumptions, instructional methodologies, technologies, clients and cultures to the current situation and projections for the future. Although a variety of delivery systems are in use at Columbia (all of which have their advantages and disadvantages) the focus will be on the university's seven years of experience with two-way videoconferencing.

Background

Before addressing the various aspects of teaching via two-way video, it would be useful to set the evolution of Columbia's distance learning programme in context.

Columbia University's School of Engineering and Applied Science created the Off-Campus Graduate Education Program and the Columbia Video Network (CVN) in 1986 to serve the needs of engineering professionals by providing graduate degree programs in several disciplines. The programme is based on a partnership with industry; students are supported by their employer both financially and administratively. Students can earn the Master of Science degree, Professional degree, and take courses toward the Doctoral degree without having to leave the work site. The programme resides within the School of Engineering and Applied Science and all current course offerings are in engineering; however, plans exist to offer programs from other schools within the University.

Initial interest in organising a graduate programme for engineering professionals came from the companies that still account for a large portion of Columbia's off-campus student population. At a meeting in

1985, senior technical managers from IBM, and faculty and administrators from Columbia identified several academic areas in which the two organisations could co-operate to provide instruction leading to the Master of Science degree. This activity began with professors of materials science travelling to teach at IBM's facility in East Fishkill, New York, which is about 120 miles from campus, in the spring semester of 1986.

At the same time, discussions with senior management at AT&T quickly led to the establishment of two more sites from which students would participate, so that by the autumn semester of 1986, the Columbia Video Network transmitted its first class from the Morningside Heights campus in Manhattan to AT&T Bell Laboratories sites in Holmdel and Murray Hill, New Jersey.

Since then, over 5000 hours of graduate credit instruction have been transmitted to professional engineers at their work sites, and a growing number of students are earning graduate and post-graduate degrees through the CVN.

The video environment

Courses to be transmitted over CVN are held in specially equipped video classrooms. Three cameras capture the lectures, including an overhead camera for recording and transmitting prepared notes, models, diagrams, and other instructional support materials placed on the podium. Computer output can also be fed into the system. Undergraduate students are employed to control the cameras and audio systems remotely from control rooms adjacent to the classrooms. This configuration allows the instructor to focus on the material and on the students (both within the room and off campus), giving only occasional directions to the operator.

Off-campus students participate in regularly scheduled Columbia courses through one of two means. The first method involves a live, two-way video link using digital (switched 56 kbit/s) telephone lines. With the second method, known as Tutored Videotape Instruction (TVI), tapes are sent via express mail at the end of each class day. Columbia's off-campus students meet the same academic requirements, take the same exams, and earn the same degrees as their counterparts on campus.

This mix of delivery systems can be difficult to manage for instructors and teaching assistants, both pedagogically and logistically. Different methods and time scales must be employed to involve students participating through a variety of means: those seated in the campus class, those connected to the class via two-way video, and those receiving the class on videotape. The presence of an on-campus class section is important to Columbia faculty and often of substantial benefit to the TVI students – their counterparts in the campus classroom provide the instructor with the feedback and questions which otherwise would be delayed and asynchronous. But the degree to which an instructor focuses on in-class students must now be balanced with an interest in two other groups of students; one group disadvantaged by the limitations of two-way video, the other without any means of 'chiming in' at all, at least not during the normal class period. TVI students present a further problem for instructors who give assignments involving a presentation, since such students are not always equipped to so.

The challenge presented by this increasingly complex environment is to create a balance between the 'no-surprises' approach to course design required for students who are unable to interact with the instructor during class; and the instructors' desire to base decisions on student interaction – decisions which ultimately affect the direction of the course, lectures, assignments, and exams.

The latter is most evident with newly developed courses, which in engineering are often quite desirable to professionals. In fact, the potential for technological innovations to make their way quickly from the laboratory to the classroom is one of the most appealing features about programs originating from a major research institution like Columbia.

TVI

Students at sites with videoconferencing equipment participate aurally and visually in the on-campus class. TVI sites receive class tapes the next business morning, and students view them with their local group at the earliest convenient time. Normally TVI groups are led by a tutor or facilitator who is a staffer with some background in the subject matter and is qualified to stop the taped lecture for local discussion and clarification. In both cases, handouts and written assignments are exchanged by express mail and electronically via the Internet. Exams are sent by special shipment to company administrators who proctor, co-ordinate communication with the instructor, and immediately ship exams back to campus for grading.

Interactive compressed video

To establish a two-way video link, off-campus sites obtain the appropriate videoconferencing equipment and access to digital phone lines, which many companies already have. A variety of codec (coder-decoder) vendors are competing to provide increasingly acceptable picture quality at 112 kbit/s, and this has become the configuration of choice for the Columbia Video Network.

Two 56 kbit/s lines, readily available from most local exchange carriers through a service known as 'Switched 56,' are joined at the codec unit. The connection can be used, as any standard phone line would be, to access the public switched network. This is significant because it enables users of compatible videoconferencing equipment to dial each other at any time and pay only for the time used, which generally costs only about twice the amount of a standard voice connection. This is in stark contrast to Columbia's original system of dedicated high-speed lines.

In addition, bridging equipment and related services have made multi-point connections possible, and costs are decreasing. All of this has persuaded Columbia to retire its original network of high-speed lines, which allowed only point-to-point communications, in favour of the new, less expensive, and more versatile technology. The last of such high-speed connections was taken down in January, 1993.

Off-campus videoconference classrooms

Corporate interest in using two-way video for workplace education has grown dramatically in response to lower equipment costs and flexible, inexpensive digital communications. Several new interactive sites have been established with the Columbia Video Network in the past year.

Most off-campus receiving sites are set up as videoconference rooms, with a conference table for eight to twelve participants. Several sites make use of existing videoconference facilities which continue to be used for other meetings and events. While this is an excellent way to test the value of interactive video for graduate education, students may occasionally find themselves competing for the space with an upper level manager who needs to hold an impromptu meeting during the semester. Once an off-campus site moves beyond the trial stage, it is, therefore, advisable for the necessary videoconference resources to be dedicated to the distance learning programme.

At minimum, these resources should include a self-contained camera and codec unit, which can be used anywhere digital lines are accessible. In this way, room assignments can be made each semester according to class size and other site-specific considerations. More extensive accommodations, such as dedicated videoconference classrooms, can be established later.

Benefits associated with the programme

Why does Columbia have a distance learning programme in engineering in the first place? As demographics affecting graduate programs in engineering and science have changed, both Columbia and its corporate participants have realised several important benefits through the off-campus programs of the Columbia Video Network:

- For the university, CVN attracts students who would not be able to attend classes otherwise and who are fully sponsored by their employers.
- Industry is able to have excellent academic programs provided without the loss of time, money, and energy for travel and logistics.
- Off-campus students benefit from these savings also, since they are able to hold a full-time job while pursuing a degree, which would otherwise be expensive and require attendance on campus.

While these monetary advantages often drive initial interest in off-campus graduate education programs, they are only the beginning of the story. Since many of the students in such programs would not otherwise be pursuing a graduate degree, it is important to recognise this more as an enabling innovation rather than a means of saving money or avoiding travel.

Columbia and its participants have also observed several other considerable benefits. Among these are:

- increased opportunities for faculty to consult and develop research relationships with industry
- increased opportunities for leading industrial scientists and engineers to serve as adjunct faculty
- continuous improvement and quality principles applied to administrative services and procedures, which benefit the on-campus community as well
- stimulating real-world engineering problems and issues being brought to the traditional classroom setting
- video segments taken in laboratories and elsewhere being integrated into the lecture
- opportunities to develop new courses, often with industry support
- industry's ability to recruit top graduates of baccalaureate programs, acclimatise them to the corporate environment, and begin productive careers sooner, and
- opportunities for off-campus students to apply what is learned to problems and opportunities at work.

Early development

In 1986, there were two primary reasons for Columbia's initial selection of interactive video as the medium of choice for its distance learning programme. Both of these offer insight into this traditional research university and how distance learning has evolved within it.

First, the general perception of the faculty and administration was that two-way video would be as close as possible to having off-campus students in the classroom. The second and related reason for selecting two-way video was the assumption that faculty would not have to change the way that they teach, since all students would have reasonable opportunity to participate in class, and most could be seen and perhaps called upon during class, regardless of location.

Both of these perceptions point to the underlying assumption that 'being in the classroom' is something to be emulated. In fact, traditional modes of classroom instruction can leave much to be desired, depending, of course, on the instructor, the material, and the students. Our current objective for distance teaching at Columbia is to change the approach to traditional instruction through the prudent use of technology. At Columbia, quality in distance learning programmes is no longer measured by similarity to the conventional classroom experience, as it appears to have been in the past.

While some early perceptions proved to be misguided, there are others which proved to be accurate. Interaction between and among the sites is facilitated by the ability to see the distant students. For example, in courses calling for student presentations, two-way video provides a nearly level playing field for on-campus and off-campus students. Without two-way video, students would have to make an aural presentation with handouts sent in advance, videotape their presentations (as TVI students sometimes do) or come to campus. Of course, the latter would be impossible for the off-campus students who are great distances from campus.

Peer-to-peer interaction among off-campus sites is also facilitated by the two-way videoconference. When students at more than one location connect to the classroom, a videoconferencing bridge is used. This means that all sites have equal opportunity to be seen by all the other sites. Students at different locations can discuss points as if they were in the room together, since their voices activate the bridging equipment to show their images. There is a slight learning curve associated with such interaction, since a half-second delay is caused by the bridging process. With practice, a protocol evolves in which the speaker pauses for replies. Off-campus sites must also keep their local systems muted; otherwise they may be seen and heard inadvertently.

Columbia's experience with all of these delivery methods indicates that one is not necessarily better than the others. Rather, a well planned mix of these and other technologies, such as computer communication and audioconferencing, can yield a learning environment that suits the professional engineer who has competing demands on his or her time.

Faculty perspectives

Over the past three years, CVN has developed a process to improve the quality of distance teaching and to expand the use of instructional technology in courses transmitted over the network.

Before discussing the various aspects of that effort, it would be useful to assess in general terms the ways in which instructors currently teach over the Columbia Video Network. While teaching is ultimately a matter of personal style, there are three general categories into which CVN faculty can be placed. These categories, and the approaches they reveal, are as much a function of academic life at a major research university as they are of distance learning systems.

Since environmental influences govern the way technology and innovation are adopted, it would be useful to create an image of the typical instructor as well as his or her approach to distance teaching via two-way video. We will focus on three broadly defined types of instructor as representatives, their attributes being based on the approximately one hundred people who have taught courses via two-way videoconference on the Columbia Video Network. To maintain a composite view and to avoid references to specific individuals, these prototypical instructors will be identified as Professors A, B, and C.

Type A

Professor A is a senior faculty member whose achievements have made him a celebrity in his field. His research projects are enormously well funded; he has written several of the most widely used texts; he consults and lectures frequently to corporations and government bodies. Professor A's interest in teaching to off-campus students hinges mainly on his goodwill toward the School of Engineering and Applied Science.

Although Professor A relies solely on the traditional lecture format in combination with weekly assignments, mid-terms, and final examinations, the material contained in his courses truly lends itself to this type of delivery. This is true of many engineering courses in which the instructor lays out concepts and problems, then shows how to solve them. Often these concepts have recently emerged from research, so instructional support materials are limited to printed images and graphs created on desktop computers.

Professor A would draw large numbers of students to his courses solely on the recognition of his name. He would not be likely to drastically restructure his pedagogical approach to meet the needs of distance learners, nor would anyone force him to do so. Fortunately, he is as serious about teaching as he is about the other aspects of his professional life, and there is really little for him to change. Students in both environments enjoy his lectures because they are so cogent and well composed that the subject's density is not only fathomable, it is stimulating. What is unfortunate about Professor A's style is that he is most comfortable lecturing from the blackboard. Aside from the textbook and occasional handouts, the only written material students receive is from the board, so note taking is a substantial part of class.

Handouts and weekly assignments are handled by CVN administrative staff, who receive all materials for off-campus students and ship them to the appropriate sites. Each off-campus site has an administrator who receives the shipments and does local routing, as well as other administrative support functions.

Professor A is quite open to using new resources, so long as they can help him meet his specific instructional objectives. He recently took up an offer made each semester by the CVN technical staff to pre-record laboratory activities, experiments, demonstrations, or to adapt existing video segments to the instructor's specifications. Professor A arranged a six minute presentation in which he and two research assistants demonstrated one of their experiments. Many students in the on-campus environment never see such examples or applications of the concepts they study, so they, too, were pleased to be taken through this window into Professor A's lab. We expect to see a steady increase in the use of this service, by a variety of instructors.

Interaction during Professor A's classes is limited mostly to the beginning of the lecture, when he entertains questions concerning the previous week's assignment. Student interaction plays out much more in the process of completing weekly assignments than in group discussions during class. This happens in at least two very traditional ways. First, there is the tangible effort a student puts forth to complete the assignment. This leads to direct interaction with Professor A (albeit asynchronous) when he reviews the student's work, which he does always, even if a teaching assistant has reviewed the work beforehand. Second, there is the peer-to-peer learning and discussion taking place in small study groups. Such groups exist on-campus, off-campus, and increasingly via electronic mail, bringing students together regardless of location. Professor A and his assistants can be reached by electronic mail as well.

Not much has changed in Professor A's approach since he began participating in the off-campus programme. It is as if a number of rooms had been added to the building in which he teaches, adjacent to the classroom, with modest-sized windows placed in between. The Professor has only to prepare material enough in advance for it to be shipped to the students 'behind the window.' For the most part those students appear to be perfectly happy with this arrangement, although more will be said about student perceptions later in the chapter.

Approximately 40% of the instructors teaching on CVN are represented by Professor A.

Type B

Professor B is an untenured assistant professor who spends most of her waking hours on research, teaching, and service to the university. Professor B is motivated to teach on the network for several reasons. First, she is aware that the administration looks favourably on this as a contribution to the well being of the school. Second, she has discovered by previous exposure to industry managers through the off-campus students that it can lead to new professional relationships and additional corporate sponsorship of her research. Third, there is additional compensation for teaching on video, which may be more meaningful to her as a junior faculty member than it is to Professor A. Although her talents are very much in demand, she has little time for consulting work with so much effort focused on the research, publications, and teaching that will become criteria in the tenure process.

Because of her teaching style and the content of her courses, she makes perhaps the best use of the interactive component of the Columbia Video Network (CVN). If her remote students could not be seen and heard, they would be more like viewers of a televised talk show than students in a graduate level course. With interactive video, their presence is actually a boon to the on-campus environment. These students bring the experience of professional engineers, real-world issues and problems of industry, and occasional challenges to the instructor's discourse, which are rarely part of the traditional classroom environment. The information added by video, which even provides glimpses of various corporate cultures, can enliven the course and provide a vista into industry for on-campus students who would not have such access otherwise.

Professor B lectures from prepared notes and graphics, which are desktop published and provided to students at the beginning of the semester in a course guide. This spiral bound document includes a complete list of the activities and time lines for the course. The course guide is part of an initiative undertaken by the CVN staff to encourage and support faculty in providing off-campus students with all of the information they need to balance educational and professional activities. As a result, students on campus have access to a valuable new instructional resource as well.

For the discussion portion of Professor B's classes, she leaves the podium and joins the students in a circle. This type of classroom setting calls for deft control of cameras and audio, and Columbia's classrooms are configured such that all of this is managed by a video technician in an adjacent control room. With planning and teamwork, everyone has an opportunity to participate, which they had better do, since a portion of their grades will depend on it.

Approximately fifty percent of the instructors teaching on CVN are represented by professor B for their general situation and approach;

however most do not leave the podium for extended discussions during class.

Type C
Professor C is an adjunct faculty member, employed full-time by the research arm of a multi-national corporation located about one hour from campus. Professor C frequently uses videoconferencing in his work. He is trained in instructional design and delivery, thanks to his corporate training department and its effort to prepare him for seminars he has given within the company. Professor C has always been exceedingly well prepared for distance teaching. In fact, the process of creating course guides was an outgrowth of his preparing materials for printing and distribution before the first day of class each semester.

Professor C's teaching style is a blend of those described above, since approximately half of class time is devoted to lecture, and the other half is a discussion of the projects that will comprise student grades. However, this workshop setting is slightly different than either of the two described previously, since Professor C's students often have materials to show to the class. On campus, this is done by going to the lectern and taking the instructor's position. Off-campus students, except for students at TVI sites, are able to use the equipment at their conference tables. Members of the latter group obviously cannot make visual presentations during class and have to rely on telephone office hours and electronic mail for interaction, which are less desirable but seem to suffice.

To support his lectures, Professor C makes frequent use of visuals such as models, videotapes, and photographs.

Professor C is motivated by a desire to teach and contribute to the future of his discipline beyond his own work and that of his company. He is also attracted to the status and prestige associated with being an adjunct faculty member of an Ivy League university.

Approximately ten percent of the instructors teaching on CVN are represented by Professor C.

Exceptions and extremes

These three prototypes represent the vast majority of instructors who have taught on the Columbia Video Network. There are of course exceptions and extremes which bear further discussion. Examples of the positive extremes include the creation of specialised software to allow electronic submission of assignments via computer, special animation to support the presentation of complex points, and personal visits to students at distance learning sites. Examples of problematic exceptions, which have become exceedingly rare, can be placed in three categories:

- *Last-minute preparation – for example creating handouts right before class*
  Although it does create an interactive learning environment, in this regard, two-way videoconferencing is no different from other distance learning delivery methods. While some videoconferencing systems allow for facsimile transmission over the same line as the video and audio, logistical issues (such as getting enough copies and the sometimes degraded quality of the facsimile image) remain a problem. When issues such as last minute preparation arise, they indicate a need for better organisation and planning, both of which are required for distance learning to succeed, regardless of delivery systems.

- *Forgetting about or ignoring the off-campus students during class*
  With the combination of a live group of students on-campus and the lack of experience interacting with people appearing on a video monitor, off-campus students can be stifled. Interaction must be planned for and encouraged early in the class.

- *Problems with the video classroom environment*
  One instructor found it nearly impossible to use the thicker railroad chalk required for the video image (normal chalk does not show up well). A more common problem for those who still rely on the blackboard, is using more than one section of the board at a time. Writing across the board is almost impossible for off-campus students to read because their view is limited by what the camera can capture.

Training

Most of these problems are things of the past because faculty members who may have experienced them receive better training. If problems do occur, faculty receive immediate coaching and support. The effort to pre-empt problems and to provide immediate assistance if they should occur are both part of CVN's continuous improvement process. The process was developed by Jane Alexander and Anne McKay, both Assistant Directors of the Columbia Video Network. It starts with individual briefings with each instructor, new or returning, who will be teaching in the following semester.

Each prospective instructor is briefed with details on administrative requirements, deadlines, and other commitments, which can change on a per-semester basis. The meetings also serve to screen faculty who may not wish or be able to meet the criteria for distance teaching on CVN. Once administrative procedures are clearly understood and agreed upon, the instructor spends time getting used to the video classroom configuration and how to use it well. Instructors receive a presentation and a supporting booklet along with an opportunity to try out the concepts and tactics by being videotaped for immediate review and feedback. The seven topics covered in the booklet are:

Getting oriented – where things are in the classroom and control room
This section contains a room diagram and details on all system components and features and how the instructor makes use of them, with examples of successful use.

Video compression – how it affects what students will see
This section describes the effects of compression and transmission and how instructors should account for them when moving about, especially when using the overhead camera to present graphics and models. Videoconferencing at 112 kbit/s provides excellent image quality, but movement on camera can be blurry, especially when large portions of the image are moving. Under the overhead camera, a hand movement or shifting of a graphic can often be distracting, so instructors are asked to avoid excessive gesturing until another camera angle is established.

The audio system – what it will and will not do
A detailed description of all audio sources and special audio considerations are given in this section. Also, a protocol for interaction is suggested, with the example of allowing some extra time for off-campus students to respond.

Interaction – getting video students to talk to you
Only the most rudimentary techniques for achieving interaction are covered in the booklet, but they are supported by the meetings with faculty. First, a strong emphasis is placed just on planning for interaction – building it into the lesson in specific ways and communicating the plan to students. Other techniques include greeting off-campus students and encouraging participation at the beginning of each class, calling students by name, looking directly into the camera on occasion, giving advance notice for question and answer periods (which should occur as early as possible in the class to avoid a drift toward passivity among off-campus students), and directing questions to particular sites and individuals.

The blackboard – and adapting it to video
This section covers the acceptable use of the blackboard as a distance learning presentation medium. The boards in the video classrooms are marked according to the aspect ratio for video, and instructors should treat each panel as a page. If writing continues in a line across the board, students off campus cannot follow the presentation. Likewise, the board should be used panel by panel, from left to right, so that references to earlier writings can be framed quickly by the camera operator. Instructors need also to consider how long students off-campus will need to take notes from the board and direct the operator to hold the shot accordingly. Finally, instructors are reminded to stand aside after writing on the board.

While this may seem obvious, new instructors sometimes find it difficult to remember.

The overhead camera – how to prepare visual materials
Since the overhead camera offers all the functionality of the blackboard, the on-campus classrooms are equipped with monitors, and pads with markers are always available at the instructor's podium. Faculty are encouraged to produce graphics and notes before hand, then to make annotations and special notes during the lecture. Faculty must consider the shape, amount of information, and colour and contrast of their visuals. Materials produced for a standard overhead projector can sometimes be used, but it is best to reformat these to account for the aspect ratio (shape) of the US. standard NTSC video screen, which is three units high by four units wide. The amount of information on a prepared page should be limited to forty-five characters in a 12 point font with no more than ten double-spaced lines per screen or page.

Clothing – what not to wear
Although it may seem trivial, the instructor's clothing can be distracting to off-campus students if the pattern is complex or contains high contrast levels. A herring bone jacket would present such problems, as would a finely striped shirt.

Feedback
Once the semester begins, each professor receives a videotape of his or her classes on a weekly basis. Tapes are also reviewed periodically by the Assistant Director for Instructional Quality, who notes any technical and pedagogical issues (positive or negative) and provides feedback as appropriate.

Faculty compensation

Faculty teaching on the Columbia Video Network receive a stipend above base salary for the additional effort of teaching on video. Part of this amount is a flat fee, to compensate for the preparation of course lectures and materials. Both faculty and teaching assistants also receive an amount for each student enrolled in the course by video, which is also provided in consideration of the extra effort to serve off-campus students.

Examination procedures

Exams at videoconference receive sites are proctored by the on-site administrative support staff. To ensure access to the on-campus proctor, remote sites follow the on-campus schedule, which is co-ordinated by CVN to avoid scheduling conflicts. Examinations are sent directly to proctors in special packets which are opened just before the scheduled exam time. The packets contain specific instructions from the faculty member and from CVN staff as necessary. At the end of the period, the completed exams are immediately photocopied and returned to campus via express mail for grading. Photocopying is done to ensure against loss or delay in the shipment of originals to campus.

Student perspectives

Given the choice, would students always select interactive video over other delivery systems? Some interesting anecdotal evidence suggests otherwise. Many students value flexibility (time shifting) over interaction. However, Columbia's students appreciate the opportunity to 'attend' class interactively, with a 'back-up' tape in case a scheduling conflict arises or for review purposes. For this reason, Columbia allows off-campus students to record the lectures and place them in a tape library for use during the semester. Tapes are also made for students on-campus and placed in the Columbia Engineering Library. At the end of the semester, all tapes are destroyed unless other arrangements are made with the instructor.

Once students become accustomed to interactive video, it can be difficult for them to accept losing the link to campus. This problem arose for one of Columbia's member companies when plans were made to shift away from the older, more expensive videoconferencing system. Before the transition to the newer system could be made, students were forced to participate in courses via videotape. These students were not pleased, and were quite open about their displeasure.

Although students' anecdotal feedback varies widely regarding the differences between interactive video and other delivery methods, no quantitative measurements have been taken on Columbia students as of yet. We plan to gather such data as part of next semester's (autumn 1993) course evaluations. A review of graded course work shows no statistically significant differences in the performance of students involved in the live interactive sections versus those receiving videotapes. For the six semesters from autumn 1990 through spring 1993, the grade point average for students receiving courses via videotape has been 3.12 or just above the letter grade of B. The average grade for students receiving courses live has been 3.17.

Analysis of student response and performance across the delivery media (conventional classroom, two-way video, one-way video with two-way audio, and videotape) indicate that, at Columbia and elsewhere, the most important aspects are a clearly communicated course plan, and sufficient student/instructor interaction. Columbia faculty are encouraged and actively supported in the development of course plans, which ideally take the form of the manual produced by Professor B described above.

Student interaction can take many forms, and we have found that a mix of electronic mail, audioconference and videoconference review sessions (also involving students who normally receive tapes, according to equipment availability), and tutoring from colleagues off-campus can combine to recreate and improve upon the traditional study hall environment.

Outlook

There are many challenges to be faced in developing distance learning programmes. The use of new tools and technologies is one such challenge, but it is a relatively simple one compared to the underlying issue of advancing new concepts in teaching and learning within the traditional Ivy League environment. As CVN has developed, there have been enhancements for students both on- and off-campus. Administrative and academic policies have been re-examined and improved, curricula have been enriched, processes have been streamlined, and a better understanding has emerged among the key participants: students, faculty, administrators and corporate managers. Concepts that are relatively new to a large research university but fostered by the off-campus programme include:

- the development of learner-focused programmes with a view toward the student as a customer
- an interest in total quality and continuous improvement
- real competition among respected institutions, independent of location (proximity to a student population) and class times (the hour at which classes meet on campus).

The future appears to be even more challenging. In addition to a growing number of competing off-campus programmes in engineering, we expect:

- the need for universities to diversify from the base of traditional graduate degree programmes and create new ones that will be shorter, more focused, and applicable at work, such as certificate programmes and short courses

- the students' ability to select the best institutions and instructors for individual graduate courses and compose programmes to suit specific objectives, which may eventually call for formal methods of assessment and accreditation to facilitate such multi-institutional degree programmes
- continued down-sizing among technology based companies, which has the double effect of reducing budgets in support of education and the technical work force of potential students
- the need to serve displaced workers who often need further education to obtain new employment.

Through the Off-Campus Graduate Education Program and the Columbia Video Network, the University has overcome the restraints of same-place, same-time delivery of traditional graduate programmes. We have also introduced innovations and new approaches to teaching and learning, which benefit all students, including those participating on campus. Over the coming years, the School of Engineering and Applied Science will apply technology to a much more ambitious strategic initiative.

The School is now in the process of opening itself in a variety of new ways. The concept is first to recreate, then re-invent, a university environment involving researchers, instructors, and students. Most or all of the activities normally associated with an academic environment and currently available only to individuals physically on the campus, can now be conducted electronically.

As they relate to instruction, such activities include upper level graduate courses and seminars, work done on doctoral dissertations, technical presentations, advising and mentoring, and other highly interactive small group activities. As they relate to research, the activities include conducting the research itself, regardless of location, co-operating with colleagues at other institutions and at corporations, reporting on research results to sponsors and peers via multimedia conferences, and many other possibilities which are as yet unforeseen.

The technological resources necessary to support these activities are either available or will be soon. Many aspects of this 'virtual academic environment' are actually occurring, if sporadically, and will be facilitated by imminent advancements in supporting technologies. The task before us as educators is to open the culture of traditional institutions so that such activities are not marginal but woven into the fabric of academia.

References

Baldwin, L. (1991). Higher education partnerships in engineering and science. *Annals of the American Academy of Political and Social Science*, **514**.

Keegan, D. (1986). *The Foundations of Distance Education*. Croom Helm, London.

Moore, M. (1989). Three types of interaction. *American Journal of Distance Education*, **3**, 2, 1-6.

Rogers, J. (1977). *Adults Learning*. Open University Press, Milton Keynes.

Rossman, P. (1992). *The Emerging Worldwide Electronic University*. Greenwood Press, Wesport.

Chapter 5

Videoconferencing in a multicampus setting

Patrick Purcell and Gerard Parr

Introduction

The University of Ulster, because of its layout (four dispersed campuses in the province) and its special mission (educational outreach to the local community) has had a long standing and serious interest in the benefits to be derived from the employment of technology supporting distance education and training in local industry. In 1990 the University commissioned one of the earliest digital videoconferencing systems in a distributed university environment, joining three of its four campuses, namely Jordanstown, Coleraine and Londonderry. Figure 5.1 shows the disposition of the sites. The Londonderry campus is 32 miles from Coleraine and 72 miles from Jordanstown. The system is intensively used for the academic programme of lectures and seminars. The system is also used by academic and administrative staff for meetings and other university business. In the context of videoconferencing usage, the University has carried out a sustained user oriented programme of monitoring the impact of the system on its academic programme both from the tutor and the learner perspectives, manifested in two reports carried out in 1992 and 1993 (Dallat *et al*, 1992 and Dallat *et al*, 1993)

The Ulster videoconferencing system is deployed with a H.120 codec located in each of the three campus sites, joined by a 2 Mbit/s line. An MCU (multipoint bridge) links the sites in any desired pattern. The lecturer sits at a desk facing the class at the local site and can see the distant site on a small monitor on his/her desk. The lecturer can also communicate with the distant site via a camera at the back of the distant class room, which the tutor can control remotely for zoom or focus by using a touch sensitive tablet and stylus at the console. This sends a signal to a Tandon-386 cue computer at the master site. Each tablet has a seating plan of the videoconferencing suites.

The lecturer's graphical support (in the shape of diagrams, pictures, headings etc.) appears as still images on a second monitor in each suite.

Figure 5.1   *Disposition of campuses: University of Ulster*

A prime rationale for installing this system was to ensure the availability of important courses to a dispersed student population with a thinly distributed demand. The appraisals referred to here, relate to the delivery of courses, which, in the main had been developed for traditional media and delivery.

The appraisal of the videoconferencing system was a formal exercise. It included both internal evaluation by the course tutors and external evaluation by appointed research staff. The methods adopted included interviews with lecturers and students, attendance by evaluators at course sessions, and the completion of questionnaires. It also included access to and analysis of tutors' videoconferencing files, notes and diaries throughout the course of the academic year.

The program of successive appraisal provides feed back to successive semesters and to planned future extensions and upgrades of the technology. The studies looked at issues such as technostress in the tutor population, the nature of the impact on the tutors' pedagogic styles. They also considered specific concrete issues such as monitor legibility, the alignment of cameras with monitors to enhance tutor/student communication and the contribution that in-session hardcopy facilities (such as a fax machine) would have on education delivery.

The studies suggest that the use of videoconferencing techniques challenges the teacher more than the learner, also that each course delivered by videoconferencing methods should include some measure of face-to-face contact. In general, it was thought that videoconferencing had a positive role to play in the delivery of educational courses, confirmation of which lies in the recent decision for proposed extensions in the use of videoconferencing in the University. It was also found that the employment of videoconferencing techniques encouraged students to become more responsible for their own learning patterns, also promoted group cohesiveness, group independence of the tutors and flexible patterns of learning. The University's programme of appraisal and evaluation continues and during the course of the next year will evaluate the effectiveness of videoconferencing as a vehicle for the delivery of a postgraduate course of study .

The University's most recent innovation in videoconferencing has been in the area of narrow-band ISDN (Integrated Services Digital Network). The opportunity to use ISDN to access a new course offering transmitted by LIVE-NET, (the London Interactive Video Educational NETwork) presented itself in January 1993. For the University of Ulster the transmission vehicle was a new 128 kbit/s videoconferencing product from BT plc – the VC7000.

The HCI module

The new course referred to above was a module in HCI (Human Computer Interaction) specially developed jointly by University College London and Imperial College of Science and Technology for videoconference delivery (Sasse and Finkelstein, 1992). This course aimed to graduate students with the necessary skills and experience to design, implement and evaluate human interfaces to interactive computer systems. The rationale for including such HCI modules in the Computer Science syllabus, recognises the paramount importance of the user and the user's needs in the development of today's computer applications. Human Computer Interaction as a academic topic of study and research is a relative newcomer to the academic scene, having its origins in the nineteen sixties, when initial efforts were made to improve the usability of computers. The progress of HCI as a subject was accelerated with the advent of the personal computer and the concomitant rapid growth of a non-professional user population. The subject matter of HCI is inherently interdisciplinary, drawing on psychology, ergonomics (human factors) and computer science. More recently, disciplines such as sociology, linguistics and anthropology have contributed to the syllabus of HCI. The course was assembled as a series of nineteen lectures starting in January '93 and extending over the following ten weeks of the Spring Semester. The lectures were given alternately by UCL and Imperial College with some input from the University of Ulster. Each of the three participating institutions organised their own follow up tutorials, seminars and revision sessions. They also organised their own assignments, tests and end-of-term examinations.

LIVE-NET

LIVE-NET began in 1985, when BT agreed to install an experimental interactive video network to serve the federated colleges of University of London. The network (Kirstein and Beckwith, 1991) was designed to provide a pair of fibre optic channels to each of the seven sites in the University, including Imperial College and University College London. Each fibre is terminated at a central switch housed at Senate House of the University of London. LIVE-NET is designed to be highly interactive, with every site operating in simultaneous vision and sound (see also Chapter 13). With four output channels available, up to five sites can take part in a fully interconnected session, with each site appearing as a window within four quadrants on each site monitor. LIVE-NET has had a remarkable track record in the introduction of innovations, in support of lectures, seminars, training and meetings between the various federated colleges in London. The system has also been interconnected experimentally with public broadcasting facilities to participate in televised conferences at a global scale. While the University of Ulster (Magee) is not part of LIVE-NET, it was possible to become a 'site' by connecting into the LIVE-NET system over ISDN-2, which was accessible in Northern Ireland, as part of the EC STAR initiative (Special Telecommunications Action for Regional Development) (STAR Initiative, 1986).

The STAR initiative

The need for an adequate telecommunications infrastructure for Northern Ireland, a region with uneven distribution of economic activity and corresponding variations in population density and industrial activity profile, became very prominent in the mid to late eighties. This requirement was further compounded with the fact of being located on the periphery of the European Community. The needs were addressed by the STAR programme. This programme was unique in that the level of telecommunications provision deployed by BT, was substantially in advance of the actual current business requirements in the greater part of Northern Ireland. Indeed, during September 1992, Europe's first SDH (Synchronous Digital Hierarchy) 3-Ring system was commissioned in Northern Ireland, with capacity to support 155 Mbit/s (Parr and Wright, 1992). The STAR rationale was, and still is, based on the conviction that peripheral regions and rural locations within these regions must have adequate state-of-the-art digital communications services to compete successfully across Europe and beyond.

Hence, the communication service adopted by the Ulster-LIVE-NET pilot exercise, was ISDN-2 with VC7000 units at both Magee College and the London end (see Figure 5.2). At the LIVE-NET end, the audio and video input/outputs were interfaced to the external microphone and cameras of the Senate House videoconference studio. The resultant transmission quality considerably improved (STAR Initiative, 1986)

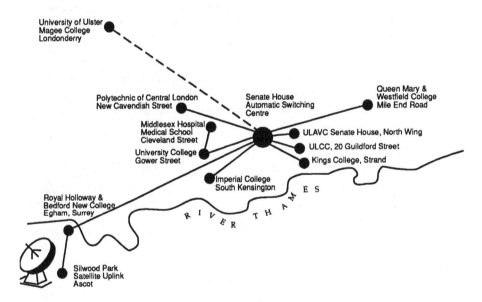

Figure 5.2   *ISDN connection of University of Ulster (Magee College) with London LIVE-NET*

ISDN: defining the terms

The conversion of telecommunications networks to digital transmission and digital switching is well under way. Economic pressures and advances in semiconductor technology have led to the increasing use of digital technology in Public Telephone Networks. Today digital techniques are being used not only in inter-exchange transmission but also in subscriber loops by incorporating analogue to digital converters into telephone handsets (hence higher quality and faster connections), and transmitting the data over 64 kbit/s digital channels.

However, end-to-end digitisation allows the standard 64 kbit/s channels employed throughout digital telephone networks to be used not only for encoded speech but also for a wide range of new and existing non-voice services. The further evolution of the integrated digital network (IDN) is combining the coverage supplied by the geographically extensive telephone network with the data carrying capacity of digital data networks. This leads to the concept of ISDN.

Integration occurs in the simultaneous handling of digitised voice and a variety of data traffic on the same digital links and by the same digital exchanges. The key to ISDN is the small additional cost for offering data services on the digital telephone network, whilst incurring no cost or performance penalty for the voice services already carried on the IDN.

Interface structure

The interface structure defines the maximum digital information-carrying capacity across a physical interface. The allocation of this capacity is accomplished using channels. Channels are classified by channel types (e.g. bearer and signalling channels).

Bearer channel

Commonly called a 'B' channel, the bearer channel is a 64kbit/s channel which conveys customer information streams from one end point to another. The distinguishing feature is that it does not carry signalling information for circuit switching by the ISDN. Examples of user information streams include digital voice, data, text, video and facsimile.

Signalling channel

Commonly called 'D" (for Delta), the signalling channel conveys control information for circuit switching by the ISDN. Enhanced signalling capabilities are offered to the ISDN services because the transfer of user and network signalling information does not interfere with the transport of user data on the B-channel.

In addition to carrying signalling information, the D-channel might optionally be used for transport of slow-speed data and for telemetry information (see Figure 5.3).

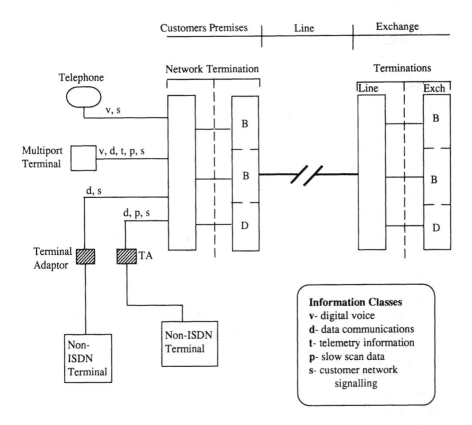

Figure 5.3   *ISDN User Access*

Due to the larger data rates which result from the use of digital technology, a channel structure of two B-channels and one D-channel is made available at the ISDN user network interface by existing two-wire subscriber links.

Basic Rate and Primary Rate services

ISDN services are divided into Basic rate and Primary Rate, each having a different combination of B-channels and D-channels.

Basic Rate (2B + D)

Basic Rate access is over two 64 kbit/s B-channels and one 16 kbit/s D-channel plus 16 kbit/s for framing and control. It is presented to the customer at a standard interface known as I.420. Basic Rate (BR) was designed to be carried over the conventional copper pair normally used for a single analogue telephone line, but it can of course be transported via other media such as optical fibres or radio. The Basic Rate launched by BT on System-X is called ISDN-2 and conforms to the international customer interface standard (2B+D I.420).

Primary Rate (30B+D)

Primary Rate (PR) service offers 30 x 64 kbit/s B-channels and one 64 kbit/s D-channel (except in North America where the digital system hierarchy offers 23B + D). It is principally used to interconnect digital integrated services private branch exchanges (ISPBXs) over 2 Mbit/s links to digital local exchanges. The internationally conformant signalling presentation for Primary Rate is I.421, and currently the BT service called ISDN-30 is adopting this standard. The previous proprietary signalling system called Digital Access Signalling System No. 2 (DASS2) was adopted by ISPBX suppliers some time ago and accounted for BT's ability to claim a world first for the launch of a commercial Primary Rate service. BT will provide a full I.421 presentation in due course.

Customer Premises Equipment CPE

ISDN relies not just on a network functionality but also on the availability of CPE and applications. This is an area which is still in its infancy, but numerous developments have already emerged, for example, Terminal Adaptors (TAs) and digital telephone sets. TAs will be widespread in the early days of BR-ISDN as they allow existing terminal equipment to use the I.420 interface. As many TAs have a handset for voice capability it can become a matter of choice whether a unit is called a TA with a handset or a digital telephone with a TA.

Videotelephony and Videoconferencing

Terminals are already available which provide good quality vision for conferencing purposes. As costs of the terminal equipment reduce with volume this should be a major growth area. A typical application would require the two Basic Rate B-channels (equivalent to 128 kbit/s from which 16 kbit/s is allocated for speech with the remainder used for video information). The terminal equipment, adopted as part of the pilot exercise being reported here, was the VC7000 videophone.

The VC7000 videophone

During 1991 several European telecommunications operators accepted an invitation to participate in a collaborative programme on the development of products and services for the videophone market. Six of the Telcos, namely BT plc, Deutsche Bundespost Telekom, France Telecom, Norwegian Telecom, PPT Telecom Netherlands, SIP Italy – took up the invitation, becoming signatories to a five year memorandum of understanding on the European Videotelephone (EV) programme.

The main aim of the EV programme is the introduction of a pan-European videotelephone service. The technology is relatively new and even though there are CCITT standards, in the guise of Recommendation H.320 relating to narrow-band visual telephone systems and Customer Premises Equipment, there has been limited widespread experience with the available technology. Work continued within the group to develop a series of videophones that would conform to the CCITT standards and was co-ordinated from BT laboratories.

Several videophones were designed and interworking tests between them were conducted successfully. One such unit is that developed by the Norwegian company Tandberg Telecom, which has been adopted for use in the UK by BT (Seabrook, 1993) (see also Chapter 7). As a result of further recommendations by the BT/EV team, the Tandberg kit has now been enhanced and recently launched as a BT portfolio product, namely the VC7000 compact videoconferencing system.

The VC7000 unit conforms to Recommendation H.320 which embraced the better known Recommendation H.261 relating to video encoding algorithms. The videophone is digital and designed to operate across ISDN-2 which is being deployed as a service throughout the European member states. As stated in the previous section, an ISDN-2 line provides two B-channels which may be used together or separately. Whilst the VC7000 will operate at either B or 2B as its default, selection was made to have all units work to the latter rate. Other restrictions relating to picture resolution and audio coding were made.

Picture resolution

The spatial resolution of a video picture consists of a number of pixels (picture elements). Common Intermediate Formal (CIF) is a resolution originally defined to allow the transfer of video between NTSC, the technique used in the USA, and PAL, the format used in the UK. CIF pictures employ 376 pixels/line with 288 lines/picture.

A reduced resolution termed Quarter-CIF (QCIF) has found much application in low bit rate video coding. The trade-off is between spatial resolution (sharpness of detail in a picture) and the temporal resolution (number of frames per second, thus the tracking of the motion). In situations with low motion content (such as typical videophone scenarios) CIF resolution is often found to be superior, although many manufacturers have developed QCIF as the default to give better rendition of motion.

Audio coding

As indicated above, a transmission rate of 2B was adopted by the VC7000. On initiating a call, the videophone units exchange capabilities – this is more or less a handshaking process whereby the units inform each other of their functionality and defaults settings. The units then switch to the highest mutually acceptable mode of operation (and hopefully performance). This is generally taken to be CIF video coding with high quality audio. In the case of two VC7000s calling each other they will default to 16kbits/s for audio, leaving the rest for higher quality video.

Additional features exist relating to hands-free audio, the ability to interface external and higher quality cameras and monitors (including rostrum cameras for graphics or documents), freeze picture facility, audio mute etc.

Experimental phase and considerations

While the VC7000 is represented as 'feature rich', there were a number of sought for features which the Ulster-LIVE-NET collaboration brought out. Several issues addressed the performance of the on-board microphone in the VC7000 and the resolution of the VC7000 camera unit. It should be borne in mind that educational videoconferencing applications were not amongst the original VC7000 applications. Also the lecturers visual aids (overheads and the diagrammatic material) being recorded by the external rostrum camera had to be held rigidly as the transmission rate resulted in appreciable times being taken to handle overhead slide movements and such like by the lecturer. Nevertheless, the exercise between Ulster and the London colleges, to transmit the HCI module over a full academic term, was successfully concluded at the end of the Spring semester.

For the purpose of the LIVE-NET exercise, Panasonic Camcorders (with both guided and free space connections) were connected to the VC7000, via a master control panel. In addition, for flexibility, and in particular for accepting questions from the floor of the lecture theatre, a radio microphone was used at the University of Ulster. It was also possible to expand the functionality of the basic VC7000 unit, by taking the audio output and feeding it into the public address system of the lecture theatre. In addition, the video was projected to a back-projector which in turn displayed it on a 10 x 6 ft screen at the front of the lecture theatre (see Figures 5.4 and 5.5).

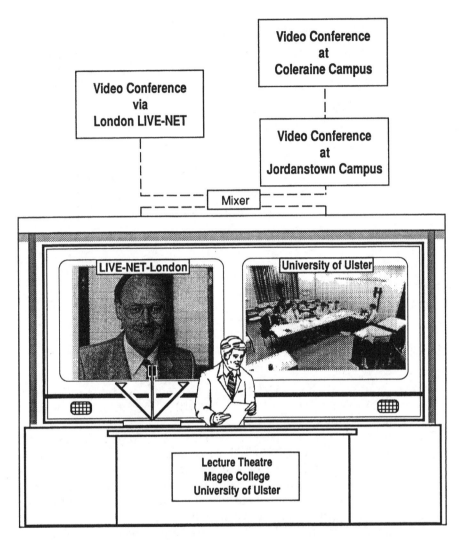

Figure 5.4 *Schematic of lecture theatre layout and videoconferencing connections*

As may be gleaned from this paper, the investment by the University of Ulster in digital videoconferencing networks as a delivery mechanism continues to be very significant. It may be said that the University has investigated the full spectrum of available bandwidth from 64 kbit/s to 2 Mbit/s and beyond.

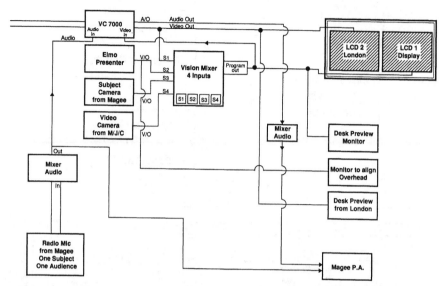

Figure 5.5   *Block diagram showing integration of VC7000 unit with University of Ulster's videoconferencing network*

The University is also exploiting its experience of educational videoconferencing in various ways.

- In the context of the DAVIDE (Digital Audio Video In Distance Education) project (Parr and Purcell, 1992), it intends to expand its links with colleges in the London area.
- The University is also a partner in the TIRONET (Trans Ireland Optical NETwork) project. The TIRONET project is based on an advanced optical fibre network linking Belfast and Dublin and is funded by the EC RACE (R&D in Advanced Communications Technologies in Europe) programme. The primary goal of this programme is to establish the interworking of optical fibre networks between EC member states.

  The academic partners in TIRONET, apart from University of Ulster are Queens Belfast and University College Dublin. All the applications which form part of the TIRONET project have relevance to the future of educational videoconferencing on the university campus. They include sharing access to desk videoconferencing facilities and to multimedia educational materials, access to shared library resources, and shared access to supercomputer facilities.
- Recently plans have gone forward to exploit forthcoming access to high speed networks such as SuperJANET as a facility for educational videoconferencing in the topics of psychology, human computer interaction and some elements of computer science.

The future development of digital videoconferencing represents a subtle mix of demand pull and technology push. Presently, the prevailing criterion for investment is to establish the economic justification. This criterion is succinctly articulated in the recent funding initiative of the UK Universities Funding Council "to make teaching and learning more productive and efficient by harnessing modern technology" (Dallat *et al*, 1993). The question of the day, is how to service the current demand to extend educational services to an ever increasing number of students against a background of progressive year-on-year cuts.

In Ulster, the continuous investment by the University in the technology of videoconferencing represents a significant ongoing commitment, if the discharge of its mission to provide the maximum number of educational options across four dispersed campuses is to be fulfilled. Current evidence of this commitment is confirmed by recent decisions to incorporate the fourth campus, Belfast into the University's videoconferencing network.

Conclusion

The recent pilot project with LIVE-NET provided conclusive evidence of the role of ISDN as a facility to extend the reach of educational delivery, not only to other campuses and universities, but as a basis for outreach to the community. It also indicated the way in which specialist skills and experience may be shared on a flexible and cost effective basis with other universities.

It can be said, in the main, that educational videoconferencing largely employs familiar techniques of pedagogy, 'the chalk and talk' of the traditional paradigms of educational delivery. Refinements in the technology such as adaptive audio systems and robotic cameras are still largely in the development stage. Furthermore, the distinct advantages that digital videoconferencing has to offer over its analogue predecessor, are neither being fully understood nor being effectively utilised. Digital videoconferencing is largely viewed by producers, programmers, and editors alike, as just a new form of analogue video, giving us a transitional 'horseless carriage' period, rather than making a clean break to establish a new basis for the future of video design and development. The radically different and more powerful features that digital videoconferencing has to offer in the areas of production, storage, reception and transmission, are but slowly being acknowledged (Purcell, 1992).

For example, in digital video productions, a much extended range of user interaction techniques are available. Digital video at the receiver offers unprecedented flexibility in the wide range of reception options, including frame rate independence, aspect ratio independence, format, resolution, bandwidth and screen size independence. Technical advances in digital storage have introduced the concept of 'dial up' movies and video.

Digital videoconference transmission is often discussed in terms of alternate delivery channels, either via computer networks, terrestrial broadcast, cable, satellite broadcast or telecommunication delivery. In fact, the future availability of digital video should result in a plurality of concurrent options, where the determinant for delivery of videoconferencing will simply be the location of the user and the convenience of the user, irrespective of whether the user is in the home, classroom or office. In this scenario educational videoconferencing is proposed as one of a range of future value-added information services together with other tele-working and leisure options.

Current advances in the technology of educational videoconferencing abound, particularly in the areas of image compression, information processing, digital storage, available bandwidth. However, the most important pointers to success will be determined by the future levels of usability and user acceptance. Recognition of this factor underlies much of the rationale of the University's ongoing user appraisal studies. Let us keep in mind the words of the educationalist Prof. A. W. Bates (1991), "I want educators rather than technical idealists to be in the driving seat, and above all I want to make sure that learners are not run over by the technology."

Acknowledgements

The authors wish to acknowledge the substantial contribution made to the subject matter of this paper by their colleagues in the University of Ulster, and at UCL and Imperial College. They include Deirdre Terrins-Rudge, David Ruddick, Nigel Gardner, Mike Jones, Angela Sasse and Anthony Finkelstein. Last, but by no means least the authors wish to pay tribute to the staff of the LIVE-NET project in the University of London, without whose unstinting co-operation the pilot exercise would never have taken place, notably the Director, Richard Beckwith and Sally Perry.

References:

Bates, A. W. (1991). Third generation distance education: the challenge of new technology. *Research in Distance Education*, **3**, 2, 10-15.
Dallat, J., Fraser, G., Livingston, R., and Robinson, A. (1992). Videoconferencing and the adult learner. Univ of Ulster Report, Coleraine, Northern Ireland.

Dallat, J., Abbot, L., Livingston, R., and Robinson, A. (1993). Videoconferencing and distance learning. Univ of Ulster Report, Coleraine, Northern Ireland.

Kirstein, P. and Beckwith, R. (1991). Experiences with the University of London interactive network. *Electronics and Communication Engineering Journal*, Feb 1991.

Parr, G. and Purcell, P. (1992). DAVIDE – digital audio video in distance education. Internal paper, University of Ulster, Londonderry Northern Ireland.

Parr, G. and Wright, S. (1992). The convergence of high-speed networking services: A report. *BCS NI Yearbook*.

Purcell, P. (1992). Broadcast graphics. In: *STAR Report, Eurographics '92*. Cambridge Univ., UK. (with Neil Wiseman).

Ruddick, D. (1993). Videoconferencing in the University of Ulster. In: *Ericsson Ireland Journal*.

Sasse, A. and Finkelstein, A. (1992). Human computer interaction: course notes. London, Dec. 1992.

Seabrook, G. (1993). Towards the implementation of a European videophone service. *BT Journal*, **11**, 1, January.

STAR Initiative (1986). European Community Council regulation 3300/86. Brussels, October 1986.

Chapter 6

ISDN-based videoconferencing in Australian tertiary education

Colin Latchem, John Mitchell and Roger Atkinson

Introduction

Given Australia's vast distances between cities, institutions, campuses and rural and remote centres, it is not surprising that Australian tertiary institutions are making growing use of videoconferencing. By July 1993, 22 of Australia's 35 public universities had videoconferencing facilities at 51 sites and most of the remaining universities were at some stage of planning to acquire systems. A further 20 sites were operational in the technical and further education (TAFE) sector with plans for further development in several states. The number of tertiary education sites may therefore easily become 80–100 in a short period of time. The most common form of videoconferencing involves two-way audio and two-way video transmitted as compressed digital signals via ISDN at 128 kbit/s.

In July–September 1992, a survey team was commissioned by the Commonwealth Department of Employment, Education and Training (DEET) to inquire into the adoption and use of videoconferencing in higher education in Australia and to make recommendations about the future potential of this technology in the light of Australian and overseas experience. The resultant report (Mitchell *et al*, 1992), which was part of a DEET Review of Modes of Delivery in Higher Education, is summarised and updated in this chapter.

Trials of ISDN videoconferencing in Australia

During 1990, Karratha College and its branches at Tom Price and Paraburdoo in the north-west of Western Australia and the Central Metropolitan College of TAFE in the WA state capital Perth, initiated LIVE-NET, a satellite trial of two-way video, two-way audio at 384 kbit/s using CLI Rembrandt codecs at four sites (Davy, 1990).

LIVE-NET continued into 1991, together with an innovative new development, TES-Sat, which utilised a mobile transceiver to provide a wide range of programs in remote and rural centres throughout Western Australia.

Terrestrial trials of inter-campus compressed digital videoconferencing were first carried out by the University of New England (UNE) in New South Wales in 1989 and the South Australian Department of Employment and Technical and Further Education (SA DETAFE) in 1990. At a time of restructuring and mergers within the Australian higher education system leading to new, multi-campus institutions, the UNE/Telecom trial of Megalink-based 2 Mbit/s and 384 kbit/s videoconferencing between the Armidale and Coffs Harbour campuses was extremely timely. Economics staff at Armidale were persuaded to use videoconferencing for all lectures and tutorials in a two-week teaching block at Coffs Harbour, 220 km to the east. The trial involved internal students enrolled at Coffs Harbour and some external students.

The evaluation concluded that for internal students, the technology was not equally effective over different teaching and learning initiatives and that interactive video technology was differentially received depending upon the teaching mode. External students expressed their satisfaction with the medium. Staff reactions were varied and, in some respects, guarded. It was concluded that there was need for an academic rationale which would indicate how this new technology could be articulated into the overall teaching program. It was also observed that the medium placed new, additional demands upon staff and that this needed to be explicitly taken into account in adopting and developing videoconferencing systems. The researchers concluded that the effectiveness of this mode of delivery derived as much from aspects of instructional presentation and communication as from the technology itself (Arger *et al*, 1989). Encouraged by the potential of this technology to assist in the work of a multi-campus university, UNE mounted a successful bid for a major grant from DEET for videoconferencing facilities on its four campuses.

The 1990 South Australia DETAFE trials involved four sites: Adelaide College of TAFE and TAFE campuses at Barossa Valley (75 km from Adelaide), Clare (142 km from Adelaide) and Gawler (42 km from Adelaide). The trials were initially conducted at 2 Mbit/s, later reduced to 384 kbit/s. They involved the delivery of 26 courses for up to 25 hours per week. Some of these courses – for example, Customer Service and Sales Techniques, Building Practice, First-line Management, Rural Property Planning and French for Winemakers – were totally new and only made possible by the advent of this new technology. In most courses, videoconferencing sessions were supplemented by distance education packages, audioconferencing and face-to-face tutorials. The medium was also used for staff development programs, conferences, demonstrations and meetings (Mitchell, 1991). The success of these trials led to widespread interest across Australia and the launch of the DETAFE Channel Network in 1991. The trials and subsequent network development entailed collaboration between SA DETAFE and Statelink (the business unit of the South Australia Department of Services) and were funded through DEET's Innovative Rural Education and Training Program. The eleven-site DETAFE Network now operates at 128 kbit/s and includes voice activated switching systems for simultaneous teaching to classes at two or more sites.

ISDN access and types of site

From 1990 onwards, compressed digital two-way video and audio dial-up ISDN videoconferencing became the standard form of videoconferencing for the Australian tertiary education sector. Initial developments centred upon 384 kbit/s bit rate using six ISDN channels plus a seventh for channel synchronisation signals. However, 128 kbit/s videoconferencing soon attracted attention with demonstrations being given at Bond University, Queensland in September, 1990 (Lundin, 1991). It was shown that the advances in codec technology were such that if 128 kbit/s was adopted as the operational norm, there was little loss in video or audio quality but considerable savings in ISDN equipment and connect time costs. In the same year, Telecom (now the Australian and Overseas Telecommunication Corporation or AOTC) completed its beta testing of *Microlink* and this form of access became available at major exchanges.

AOTC markets two types of access to ISDN. *Macrolink* provides a 2 Mbit/s capacity divided into 30 B (user data) x 64 kbit/s and 1 D (64 kbit/s signalling) channels, with a minimum subscription of 20 channels. *Microlink* provides two B x 64 kbit/s and one D (16 kbit/s signalling). Macrolink requires the customer to acquire approved equipment for ISDN multiplexing and for conversion of data to ISDN protocols. With Microlink, the ISDN multiplexing or 'mux' function is undertaken at the customer's local AOTC exchange by means of a B-mux. This represents a considerable saving for customers whose traffic does not warrant a Macrolink. Microlink customers have to install terminal adaptors for conversion of data to ISDN protocols but these are relatively inexpensive at A$2400 to A$3500 per pair of channels.

Australia has made rapid advances in installing ISDN – but in such a huge country, it is inevitable that there are financial, technical and operational limits to the geographic availability of ISDN. Microlink is only available in metropolitan locations, major country centres and further locations where significant demand exists and is dependent upon AOTC's schedules for upgrades to exchanges. Also, for technical reasons, the Microlink service is currently only available to within about 3.5 km of an exchange equipped with a mux. Signal repeaters have yet to prove their effectiveness. Therefore, for example, Curtin University in Western Australia has three Microlink-connected campuses while the fourth campus – the Institute of Agriculture at Muresk, 11 km beyond Northam – is connected by Macrolink.

A 128 kbit/s videoconferencing facility using Microlink for ISDN access is significantly less expensive and less complex than a 384 kbit/s facility. Call charges are only a third of the cost and Microlink rental costs are little more than for two telephone lines. This opens up the possibility of district sites being equipped with Microlink and rostered for portable codec use, as trialled for example, by the Great Southern Regional College of TAFE in Western Australia.

Tariff band		*One hour videoconferencing @ 128 kbit/s – cost in A$*
NDD 1	Local	5.20 (1.60)
NDD 2	25 km – 50 km excluding NDDI	15.40 (7.60)
NDD 3	50 – 165 km	35.20 (19.00)
NDD 4	165 – 745 km	43.90 (24.55)
NDD 5	More than 745 km	62.35 (35.65)

Charges apply to day rates (Mon-Fri 8 am – 6 pm). Night rates (in brackets) apply at all other times.

At 1 May 1993 the Australian Dollar was worth £0.45 and US$0.71

Table 6.1   *AOTC tariff rates within Australia (as at 10 March 1993)*

Codec manufacturers have adopted different and incompatible ways of undertaking digitalisation, compression and communications. Thus, the three types of videoconferencing codec at education sites in Australia – PictureTel, CLI Rembrandt and GPT – are unable to communicate with each other unless they are equipped or upgraded to use the CCITT's H.261 set of standard protocols (also referred to as the *p x 64* set of standards). As Table 6.2 shows, CLI Rembrandt and PictureTel have the market share in Australia with PictureTel increasing its market penetration to the point of becoming, essentially, the current *de facto* standard.

The dominance of proprietary protocols in Australian compressed digital videoconferencing reflects a number of factors. The tertiary institutions' major decisions on choice of system occurred at a time when the contenders had not produced their implementations of CCITT's H.261 and associated standards. Relatively few sites have since acquired H.261

capability, partly because of the additional expense, and partly because of the limited needs to communicate to sites with different codecs.

The operation of the H.261 set has been trialled in Australia for international ISDN videoconferencing, for example by Curtin University communicating via AOTC Switched Digital with a CLI system at Ngee Ann Polytechnic Singapore and a Mitsubishi system at the Commonwealth of Learning in Vancouver. The technical quality of video under H.261 operation is inferior to that when using proprietary protocols but with connectivity with 16 countries and tariffs at about A$400 per hour, the exciting possibilities of international initiatives outweigh this disadvantage.

The importance of connectivity between all sites varies considerably from one organisation to another. For example, the SA DETAFE videoconferencing network has a capacity load from its communications for teaching and other internal work and the importance of connectivity with other systems is currently of lesser importance. By contrast, Western Australia's Murdoch University has only one campus, and therefore connectivity to other institutions' sites, especially within the state, has to be of major concern.

Whilst access to a large number of sites undoubtedly affords more opportunities, it does not follow that access to every site in Australia is essential for effective uses of this medium. In practice, limits to the number of sites accessed are more likely to be determined by the number of working relationships which can be established between institutions for teaching, research, administration or other purposes, rather than incompatible equipment specifications. Sites which do have major and diverse needs for videoconferencing may overcome compatibility problems by installing a second type of codec, as, for example, at the various UNE and Charles Sturt University (CSU) campuses and two of the SA DETAFE sites.

The extent to which connectivity is a problem also depends upon the delivery modes envisaged. In Australia, the main uses to date have been for inter-campus administration or teaching and for regionalised approaches to providing students at study centres with tutorial support. Therefore, national connectivity may be a desirable but not immediately essential requirement.

Videoconferencing in Australian higher education received an enormous fillip in 1990-1991 when DEET, as part of its National Priority (Reserve) Fund initiative to improve the quality of education, granted over A$3.3 million for the establishment of facilities at CSU and UNE in New South Wales, Deakin and Monash Universities in Victoria, the University of Central Queensland, the University of South Australia, and Curtin, Edith Cowan and Murdoch Universities in Western Australia. The funds were granted for systems that facilitated inter-campus links, remote site access and 'open' and innovative delivery of higher education.

As Table 6.2 (updated from Atkinson, 1992) shows, more institutions subsequently installed videoconferencing facilities. In addition to the institutions listed in this Table, the University of Adelaide, the Australian Catholic University, La Trobe University, Macquarie University, University of Melbourne, University of Newcastle, Queensland University of Technology, Swinburne University of Technology and Victoria University of Technology propose to install videoconferencing as funding becomes available. Sydney University's veterinary science faculty's Sydney and Camden campuses can already link with Murdoch's veterinary science school in Western Australia and Massey University in New Zealand. They will soon be joined by Melbourne and Queensland Universities to form the first veterinary school collaboration of its type in the world. The Australian National University, Ballarat University College, Griffith University, James Cook University of Northern Queensland and University of Western Sydney are investigating their needs for videoconferencing.

Two exceptions to this form of ISDN networking are to be found at Edith Cowan University (ECU) in Perth and UNINET in Sydney. ECU's four metropolitan campuses use fibre optics to deliver broadband analog signals to each other and by casual hire of a television bearer to ECU's Bunbury campus, 200 km south of Perth. They are also connectable to digital technology gateways and satellite broadcasting gateways via AOTC (Grant, 1990). UNINET is a fibre optic telecommunications network linking Sydney University, Macquarie University, the University of Technology Sydney and the University of New South Wales. Currently, UNINET provides four broadcast-quality television channels into and out of each campus and access to the broader AOTC network (Cole, 1990).

In addition to the tertiary videoconferencing sites described, there are videoconferencing bureaux and demonstration locations in all Australia's state capitals, and facilities in a growing number of private organisations.

Applications and networks

Typically, equipment costs per site are in the A\$90,000 plus range, depending upon the number and type of cameras, monitors and other peripherals. The majority of sites have a single room camera supplemented by a document camera, a VCR, supplementary microphones and a facsimile and a telephone. Some have two or more room cameras, typically housed in rollabouts for greater flexibility. The facilities do not need technicians to operate them and they only require minimal keypad operation by the presenters or chairpersons.

Site type	institutions and number of campuses
PictureTel	*On site codecs* Charles Sturt University (3 campuses) Curtin University (4 campuses) Deakin University (5 campuses; 6-port bridge) Flinders University (1 PictureTel + 1 CLI for research purposes) Monash University (4 campuses) Murdoch University (1 campus) Queensland Government. TSN11 videoconferencing, including TAFE (equipment ordered for 14 sites plus an 8-port bridge) SA DETAFE (2 sites) *Gateways for PictureTel via an AOTC Television Operating Centre* Sydney and Melbourne TOCs have CLI Rembrandt and PictureTel codecs. These are accessible by ISDN and available for "back to back" operation as protocol converters between sites with incompatible codecs. *Other gateways* In 1992, Western Australian PictureTel users experimented with ISDN access to a PictureTel codec temporarily located at the Golden West Network's Bunbury WA uplink to an Optus A satellite transponder which provided a temporary gateway to statewide television broadcasting.
CLI Rembrandt	*On site codecs* Charles Sturt University (3 campuses) Northern Territory University Royal Melbourne Institute of Technology SA DETAFE (11 campuses) University Centre, Sydney (joint UNE, CSU, Wollongong facility) University of New England (4 campuses) University of South Australia (2 campuses) *Access to a Rembrandt CLI codec via an AOTC TOC* Sydney: UNINET (Sydney and Macquarie Universities, University of New South Wales and University of Technology Sydney). Perth: Edith Cowan University
GPT H.120	*On site codecs (Megalinks, not ISDN)* La Trobe University – University College of Northern Victoria – Bendigo, Mildura, Wangaratta, Shepparton and Swan Hill sites. *Other gateways* To a site with H.120 codec and satellite uplink, in Melbourne. Note that H.120 is an older set of standards for 2 Mbit/s only and is not compatible with the newer H.261 set of standards and not accessible from the ISDN switched network.

Table 6.2   *Codec types at Australian tertiary education sites (March 93)*

The applications and networks achieved across Australia in the relatively short timespan have been many and varied. In Western Australia, Curtin, Murdoch and Edith Cowan Universities and some TAFE Colleges equipped with PictureTel units, have co-operated in a State Government initiative – WestLink – to trial educational delivery via a spare 12 watt satellite transponder to remote sites with TVRO facilities. UNE has trialled language teaching at 64/128 kbit/s between Australia and Japan (Arger and Wakamatsu, 1992) by AOTC Switched Digital services and has recently entered into an agreement with Universiti Brunei Darussalam for the interchange of programs by videoconferencing.

The University of New South Wales, New England and Western Australia, who are partners in the national Co-operative Research Centre for Premium Quality Wool, are using PictureTel to teach wool science to undergraduates. The number of experts in wool science in Australia is very limited and mainly concentrated in the Commonwealth Scientific and Industrial Research Organisation who are also partners in this CRC. Videoconferencing enables these partners to provide a comprehensive and unique staff resource for a course that would otherwise have been impossible to organise.

In the heart of Sydney, the Clarence Street Centre, a shared facility of CSU, UNE and University of Wollongong, uses videoconferencing to assist each organisation's publicity, marketing, conferences, training programs and educational activities which have included music lessons featuring visiting Masters of Music.

UNINET is used for such activities as weekly joint research seminars and postgraduate lectures and tutorials between the four Sydney universities which have resulted in shared resources, reduced costs and increased student access to specialised presentations. Multi-campus institutions such as UNE, CSU, Monash and Deakin find videoconferencing to be invaluable in facilitating administrative and management processes. Deakin has installed Australia's first privately-owned 6-port PictureTel multipoint bridge which is used by other institutions across Australia and for inter-institutional and national meetings. Multipoint bridges will also be installed for statewide systems in Western Australia and Queensland.

Evaluation and conclusions

In the survey for DEET, higher education users of videoconferencing were found to be in no doubt that videoconferencing was an important addition to their institutions' repertoire of delivery technologies. It has allowed them to do some new things and to do some things differently and better. It was seen as having the potential to have the same growth path as facsimile and computers. At the same time, it was not seen as a panacea for all ills and it was accepted that the benefits may vary from context to context and from application to application.

The measure of interest in the survey may be gauged by the 100% return rate from the universities surveyed. The survey results indicate that institutions value videoconferencing because it yields cost savings (for example, in displacement of travel and accommodation costs), productivity gains (for example, reduction in staff down-time travelling between campuses) and strategic gains (for example, increased co-operation or collaboration or strong competitive advantage). Extensive travel also puts staff and vehicles at risk and distances between university campuses can be enormous. Examples of common return journeys between campuses that can now be replaced by videoconferencing are:

(CSU)	Bathurst – Wagga Wagga	680 km
(Deakin)	Geelong – Warrnambool	372 km
(Curtin)	Perth – Kalgoorlie	1194 km
(Monash)	Melbourne – Churchill	300 km
(UCQ)	Rockhampton – Mackay	680 km.

Discussion of videoconferencing needs also to include consideration of the potential of the medium to meet social justice objectives. Efforts have been made at some institutions to use videoconferencing to address the needs of particular groups such as Aborigines and those with particular disabilities. Overall, the study revealed that students' reactions are consistent with the findings of Hansford and Baker (1990) which were that the students who are distant from the lecturer are keener on the medium than those students who are at the same site as the lecturer. Schiller and Mitchell (1992) suggest that students who have experienced other forms of distance education may be more appreciative of the lecturers' efforts and more tolerant of any technological limitations than students who have only experienced classroom teaching.

There is growing use of this technology for multi-campus management and faculty meetings. Most of the interviewees were relatively tolerant of any inherent limitations in the technology and preferred to use it because they perceived that the managerial advantages outweighed any disadvantages. Research by Hansford and Baker (op cit) at UNE revealed that in general, participants perceived videoconferencing meetings to be as effective as face-to-face meetings. Lundin and Donker (1992) suggest that videoconferencing meetings may be less intense, less stressful and less exhausting than meetings by audioconference.

While universities make increasing use of videoconferencing for administration and committee work, most institutions had to struggle initially to attract lecturers to use this medium for teaching. As with any technology, the easier part was to acquire and install the videoconferencing facilities. The harder part was to have it used and used effectively. This has been the experience of many of the institutions surveyed. The study revealed that many lecturing staff feel threatened by this new medium. From induction course to initial successful use, to frequent use of the medium, can be a slow process. To overcome the hurdles to adoption, various institutions have developed appropriate induction and staff development programs. Youngblood, Mahoney and Tonkins (1991) suggest that typically non-users have concerns at the awareness, informational and personal levels. The approach they have successfully adopted at UNE has been the classic extension/marketing approach in which early adopters are targeted for the promotion of innovative practices and potential users are targeted according to their needs. It was found that if the need to use the technology is strong enough, potential users are more likely to overcome their inhibitions.

The survey results showed that lecturers value the following features of videoconferencing most highly:

- the relatively low developmental costs of this medium compared with video/tv and the ease of preparation compared with many alternative delivery modes
- the ability to tailor lessons to the needs of relatively small numbers of students
- the ability to offer small numbers of students at smaller or remote campuses a wider range of offerings than was previously possible
- the possibility of offering higher levels of psycho-social support to distance learners through real-time interaction
- the medium's rich visual dimensions – for example, the body language of all participants, the ability to transmit two and three dimensional objects via a document camera, and the ability to transmit computer images
- the major productivity gain of lecturers not having to expend time and energy on travel
- the opportunities for students to interact with their peers at other sites
- lecturers being able to maintain their knowledge and skills by teaching second and third year subjects from remote campuses
- the value-added benefits of linking a lecturer from one university to students at another
- the export potential of teaching directly to overseas locations.

Concerns that some lecturers held in regard to teaching by videoconferencing were:

- the extra preparation time required
- the intensity and concentration required
- the resultant tiredness of staff and students after sessions
- the restrictions on mobility within the classroom

- the slight blurring of motion and lack of lip synchronisation due to compression and slow transmission speed
- the lack of real face-to-face contact
- the fear of stage fright in a session.

Teaching by videoconferencing has been shown to be effective where interactivity and synchronous learning apply. It works best with teachers who like to use a conversational style and who prepare for, and encourage, voluntary contributions by all participants through tutorials, role plays, simulations, brain-storming, problem-solving and practical activities. It also works best when other media are involved in the delivery, for example print packages or videos distributed in advance of the session or material faxed during the session itself.

It is evident that many teachers find that teaching via this medium requires higher energy levels than in classroom teaching and calls for different and additional skills. Schiller (1992) suggests that sessions should be relatively short and should contain a variety of presentation techniques. There is also need to give careful consideration to group dynamics which are different from those in face-to-face teaching. This is particularly the case with voice-activated systems in multi-point mode where the dominant voice at any site determines which image is transmitted to all sites and the response time is slightly delayed by the slower transmission speed. Schiller and Mitchell (1992) suggest that videoconferencing requires different teaching methodologies from any that teachers may have used previously and that it necessitates different ways of moving, different ways of presenting information and different ways of judging the meaning of messages going in both directions.

Teaching applications usually make great and varied demands upon room layout and audiovisual presentation arrangements and this needs to be catered for in the design of videoconferencing facilities. Teachers need to work in a familiar environment and the typical design is based upon the seminar room rather than the television studio. Schiller and Mitchell (*op cit*) also suggest that it is important that privacy should be maintained during videoconferencing sessions so that lecturers and students know that they can talk as openly as they would within the four walls of the conventional classroom.

The survey found general agreement that compressed digital videoconferencing worked best when linking groups of up to 30 students at two to four sites. With larger groups it was less effective. It was also generally accepted that videoconferencing may not be cost-effective in all contexts. For example, UNE, which has linked its students with those of the University of the Air (Tokyo) to help teach foreign languages in both countries, considers that the current costs are probably prohibitive for regular videoconferencing for overseas language teaching but that as a complement to existing language courses there are many benefits, and that for business applications such delivery could be very attractive.

The survey also established that videoconferencing can provide an improved quality of education for rural and remote areas, and facilitate the development of workplace training. Some effective applications have addressed the needs of particular groups including Aboriginal Communities. It is worth noting that the Tanami and Pintubi people in the remote desert north-west of Alice Springs have their own satellite videoconferencing network (see Toyne, 1992). The technology can also facilitate inter-institutional, interstate and international collaboration and add considerably to the diversity and accessibility of advanced and specialised educational services. It can also bring together the best courses, lecturers and researchers and demonstrate best practice.

The drawbacks to, and the criticisms of, the medium also need to be acknowledged and addressed. Installation costs are still high, inevitably limiting the number of sites and the extent of access. It is easy to use the technology poorly and there is insufficient research on the use of the medium and the comparative strengths and weaknesses of videoconferencing vis-à-vis other distance education technologies to guide users and managers. The technology holds enormous potential but Hansford (1991) suggests that the traditional bastions of education and training are conservative and well organised against structural change and that if technological change impinges on the traditional face-to-face, teacher-student relationship, innovators should be prepared for a long period of guerilla warfare. The delivery mode may have all of the advantages (e.g. immediacy) but all of the disadvantages (e.g. weak instructional strategies) associated with traditional teaching. Taylor (1991) argues that there is a need to distinguish between delivery technologies and instructional design technologies. He calls for further research and evaluation of the pedagogical efficacy of specific courses in specific contexts and the types of distance education provision that can best meet the needs of Australian students in the present educational, political and cultural environment. These are important issues to be considered by practitioners and researchers.

The survey revealed that the medium is still so new that its potential outweighs its present usage but rapid development is occurring. Hence the survey team recommended that Commonwealth funding for installations, research and development be continued and that support for the efficient and effective growth of networks be provided through:

- a national strategic plan for the development of videoconferencing in the higher education sector
- government departments and telecommunications carriers establishing ways of amending current policies brought about by de-regulation to provide access and equity in rural and remote areas

- further research into the instructional design and learning aspects of videoconferencing
- investigating the potential of videoconferencing networks with countries in the Asia-Pacific region
- a national users' group concerned with staff development, evaluation, cost-effectiveness studies and educational and technological management
- a closer, more co-ordinated relationship between users and suppliers of the technology
- a national register of sites, facilities, fees for services and courses
- shared and inter-sector use of sites, networks, bridges and expertise.

The survey team concluded that the decision of the Commonwealth Government to encourage and invest in compressed digital videoconferencing for Australian universities had been a bold one and had placed Australia at the leading edge in this field.

Authors' note

This chapter is based upon a paper originally published in Davies, G. and Samways, B. (eds.), *1993 Teleteaching*, Proceedings of the IFIP TC 3 Interactive Conference on Teleteaching '93, Trondheim, Norway, 20-25 August, 1993. The authors are grateful to the International Federation for Information Processing for permission to republish this paper in an updated and extended form.

A national register of Australian PictureTel sites and contact details is available by anonymous ftp to `cleo.murdoch.edu.au` in files in the `/pub/video_conferencing` directory.

Alternatively, for more information send electronic mail to `atkinson@cleo.murdoch.edu.au`

References

Arger, G. and Wakamatsu, S. (1992). Experimental international language teaching between Japan and Australia using 64/128 kbps compressed video via ISDN. Paper presented at the Pacific Telecommunications Conference, Honolulu, Hawaii.

Arger, G., Jones, G., Smith, R., Baker, R., and Hansford, B. (1989). UNE/Telecom trial of interactive video using data compression techniques between UNE Armidale and UNE Coffs Harbour, 3-13 October, 1989. University of New England, Armidale, NSW.

Atkinson, R. (1992). The national educational communications framework: analysing the question of common technical specifications. Paper presented at Australian Society for Educational Technology Conference, *Ed Tech '92*, Adelaide, SA.

Cole, T. (1990). UNINET: Optical fibres in support of research and learning. In: *Converging technologies: Selected papers from Ed Tech '90* (J. Hedberg, J. Steele and M. Mooney, eds). Australian Society for Educational Technology, Belconnen, ACT, **37**.

Davy, G. (1990). LIVE-NET: The Pilbara videoconferencing project. In: *Open learning and new technology: Conference proceedings* (R. Atkinson and C. McBeath eds). Australian Society for Educational Technology, Perth, WA, 93-112.

Grant, M. (1990). Education enters the age of telecommunications. In: *Open learning and new technology: Conference proceedings* (R. Atkinson and C. McBeath, eds). Australian Society for Educational Technology, Perth, WA, 162-164.

Hansford, B.C. (1991). Education, training and telecommunications as a social activity. *Telecommunication Journal of Australia*, **41**, 3, 26-31.

Hansford, B.C. and Baker, R.A. (1990). Evaluation of a cross-campus interactive video teaching trial. *Distance Education*, **11**, 2, 287-307.

Lundin, R. (1991). *Australian teleconferencing directory 1991*. Queensland University of Technology, Brisbane, QLD.

Lundin, R. and Donker, A. (1992). *Queensland videoconference trial: Report of the Queensland government trial of compressed videoconferencing between Brisbane and Townsville*. Media and Information Service, Brisbane, QLD.

Mitchell, J. (1991). Videoconferencing in the SA DETAFE: using state-of-the-art technology to enhance teaching and learning. In: *National Technology in Education and Training Conference Proceedings*. Technology in Education and Training Committee, Redfern, NSW.

Mitchell, J., Atkinson, R., Bates, T., Kenworthy, B., King, B., Knight, T., Krzemionka, Z., Latchem, C., and Schiller, J. (1992). *Videoconferencing in higher education in Australia*. Unpublished report to the Review of Modes of Delivery in Higher Education in Australia. A report funded under the Evaluation and Investigations Program of the Department of Employment, Education and Training, Canberra, ACT.

Schiller, J. (1992). *Videoconferencing in South Australia: a review of the DETAFE Channel Network*. University of Newcastle, NSW.

Schiller, J. and Mitchell, J. (1992). *Interacting at a distance: staff and student perceptions of teaching and learning via videoconferencing*. Paper presented at AARE/NZARE 1992 Joint Conference, Educational Research: Discipline and Diversity, Deakin University, Geelong, VIC, 22-26.

Taylor, J. C. (1991). *Distance education and technology: a conceptual framework*. Paper prepared for the National Distance Education Conference Working Party on Education and Technology.

Toyne, P. (1992). *The Tanami network: New uses for communications in support of social links and service delivery in the remote Aboriginal communities of the Tanami.* Paper presented at the Service Delivery and Communications in the 1990s Conference, Darwin, NT.

Youngblood, P., Mahoney, M. and Tonkins, S. (1991). Issues in the implementation of a transcampus videoconferencing network. In: *Quality in Distance Education* (R. Atkinson, C. McBeath, and D. Meacham, eds). Australian and South Pacific External Studies Association Forum, Bathurst, NSW, 512-519.

Chapter 7

ISDN videotelephony in Norway

Tove Kristiansen

Introduction

The transmission of live video images over the telecommunication network is not a new phenomenon. Television has been a part of our daily lives for several decades, and videoconferences, with a transmission rate of 2 Mbit/s, have represented the opportunity for two-way visual communication for some years now. Communicating visually with one single digital telephone line, however, was not possible until very recently. With the building of a fully digitised telecommunications network throughout Europe, the videotelephone is expected to be widely used by the end of the century.

Norwegian Telecom Research (NTR) began studies in the field of image compression techniques about ten years ago. Through co-operation with the Norwegian company, Tandberg A/S, a video codec was produced in 1987 which made it possible to transmit live colour images over a single 64 kbit/s digital line. This codec has since been used in several field trials with dedicated 64 kbit/s connections and also, since 1990, via ISDN.

There were certain limitations to the quality of sound and image presentations of this first Norwegian-made video codec. Nevertheless, the quality was sufficiently good for the videotelephone to be a useful tool for various categories of users. Encouraged by the results achieved in 1987, it was decided that the videotelephone project be strengthened and continued, and in the autumn of 1991 a Norwegian-made ISDN videotelephone terminal with speech and video codecs included, was presented for the first time (Figure 7.1). This videotelephone may be used on one or two B-channels in ISDN, i.e. at 64 or 128 kbit/s transmission rates, dependent on the user's need for quality and the cost he or she can afford.

The videotelephone opens up a whole range of new communication possibilities, both for professional and private use. The Norwegian experiments thus far have been carried out in the areas of distance education, remote supervision, remote surveillance and sign language. For deaf people, the videotelephone represents a completely new opportunity for communicating in their own language, sign language, over the telecommunication network. Field trials have been carried out both with deaf children and deaf adults in Norway.

Figure 7.1   *ISDN videotelephone terminal developed by NTR and Tandberg*

Norway is a mountainous country with a scattered population. Distances are great and the regional areas are often short of qualified personnel. With the aid of the videotelephone, it becomes possible for people to receive an education at home, at work or at local study centres. It is possible to bring experts from centralised institutions more frequently and more regularly into contact with clients and educated personnel in regional areas. Thus, the videotelephone may help to overcome some of the restrictions geographical distances often place on people's opportunities for obtaining further education and adult training.

Design objectives of the ISDN videotelephone terminal

When designing new telecommunication terminals, such as the videotelephone, it is important to take into account usability factors that may be crucial to its success. In the Norwegian videotelephone project, great care has been taken to design a terminal that is simple to operate. The experiences gained in the field trials have had an important influence on the development and design of the videotelephone terminal.

The feedback from the field trials resulted in two distinct design criteria for the ISDN videotelephone:

- User-friendliness, in terms of installation and use, means that it should be as easy to use as a normal telephone.
- It must be only one piece of equipment with a minimum need for connecting cables and wires.

Consequently, the ISDN videotelephone terminal developed by NTR and Tandberg is just a single piece of equipment. The video and audio codec is an integral part of the terminal, placed at the left hand side of the monitor. The only cables required are those for the power supply and the ISDN connection. That means easy installation.

NTR has put a lot of research effort into developing a simple and easy user dialogue. The results, which were also submitted to international standardisation bodies and formed the basis for ETSI's (European Telecommunications Standards Institute) work in this area, prepared the ground for designing the videotelephone keyboard as shown in Figure 7.2.

Figure 7.2   *Keyboard of the ISDN videotelephone*

To the terminal there may also be connected external devices such as a monitor, a document camera or a computer. These facilities are important, and absolutely necessary when more than one person is using the equipment at the same time as, for example, in desktop videoconferencing and distance education.

Distance education

Norway has long traditions in distance education. Ever since the first correspondence school was established in 1914, distance education has represented an alternative means of learning for thousands of people with no or very few opportunities for ordinary schooling. Modern technology now offers new and more efficient ways of organising distance education. Of particular interest are the possibilities modern telecommunication represents for two-way communication.

Audio and/or computer conferences are often sufficient to meet the communication needs in distance education. In many cases they may even be the best solutions. In some situations, however, the need for visual communication is obvious. One example of this may be when a doctor wishes to transfer a video image from a microscope as part of distance teaching for medical students.

Even in situations where it is not an absolute requirement, visual communication clearly adds an extra, and perhaps valuable, dimension to the interaction between teacher and student. The possibility for teacher and student to see each other, seems in itself to be of great value. It brings them closer together and makes the interaction more direct and personal. In addition, visual communication gives the teacher the opportunity to present lectures and learning material in a way that is not possible in audio or computer based communication.

What are the benefits of visual communication in distance education?

- Teacher and students can see each other.
- Teacher and students can communicate with the aid of body language (to a limited extent).
- Illustrations, pictures and video images may be presented.
- Different working operations and processes may be demonstrated.
- Learning material may be presented in writing, as a substitute to using a blackboard.

Over the last few years, the use of videoconferences (2 Mbit/s) in distance education has expanded in Norway as in many other countries. Universities and colleges use videoconferences to distribute lectures to groups of students that are widely spread geographically. Teachers in high schools have the opportunity to increase their knowledge of special education with the aid of videoconferences. Workers in the fish processing industry are offered further education, and schools that are short of qualified teachers may supplement their teaching schedule by putting up distant teachers on screen.

The experience gained from using videoconferences in distance education is by and large very positive. Some people feel a little uneasy when first confronted with cameras and TV screens. In most cases, however, they become used to it after a surprisingly short period of time. Having participated in a few videoconferences, people tend to forget the technology and concentrate on the communication and learning processes.

The videotelephone, with its requirement of only one single digital telephone line, is of particular interest to distant educators. Through ISDN, visual communication will be available at a much lower price than the 2 Mbit/s videoconferences of today. At the same time, accessibility will be greatly improved. With this prospect in mind, the Norwegian Telecom Research has spent the last few years experimenting with the use of videotelephones in distance education.

Technical training at a distance

Norwegian Telecom employs a total of approximately 15000 people all over the country and substantial resources are spent each year on education and training. Most of the internal courses offered have until now been held at national training centres. To attend one of these courses, the participants have to travel and be absent from their jobs, families, social networks and other obligations. Due to these circumstances, there is rarely more than one person at a time from a local area attending the different courses. On returning to their daily work, they will have no one with whom to discuss this newly gained knowledge. Consequently, there is a danger that what was learned at the course will soon be forgotten.

Another aspect of this traditional way of running courses is the fact that it is difficult to find courses that are both relevant and offered at the right time. The subject taught may not be relevant to the participants' work until another two or six months have passed, or it may have been highly relevant if it had been taught some months earlier. In many situations this method of learning is therefore not very efficient.

As an attempt to improve the learning process, an experiment with videotelephones was initiated in the autumn of 1989. The aim of the experiment was threefold:

1 *To meet the needs of the workers in their everyday work, and at the right point in time*
The videotelephones made it possible to train whole groups of workers at the same work-place simultaneously. Thus, discussions would most likely arise and the competence level of the whole group be improved.

2 *To establish contact between groups of workers who are located at separate work-places but who perform more or less the same kind of work*
Sharing experiences in this way was thought to be stimulating and motivating for each person's daily work.

3 *To make it possible for the workers to contact skilled personnel in the central organisation whenever needed*
In this way, the workers could have direct training and supervision.

What the workers appreciated most about the experiment was the contact that was established with workers at other sites, people with whom they had previously had little contact. The third aim that workers would contact skilled personnel directly and spontaneously proved to be unrealistic and unattainable. To do this, the workers would have had to bypass their local tutors and other persons in charge at different levels in the organisation. This would have represented a fundamental break with established routines of communication within the organisation, and was completely out of the question.

This experience was a useful reminder of the fact that new means of communications do not necessarily lead to new ways of communicating. The use of new technology has to be adapted to the environment, and both economical, social, cultural and organisational aspects will have an influence on the communication pattern that can be established. Normally, organisational adjustments to new technological means of communication will only take place over long periods of time.

Training nursing students at a distance

Norwegian nursing students attend college for three years. During this time they spend several months practising at hospitals or other health care institutions. These institutions are often miles away from the school and this means that their teachers can give them only limited supervision. The teachers visit them during their stay, but there is no feasible way of following them up on a more regular basis.

By using the videotelephone, it becomes possible for the teacher and students to keep in touch on a more regular basis during the periods of practical training. In the autumn of 1989 a project called the TOVE project (from the Norwegian, 'toveis', meaning two-way) was set up in Mid-Norway to test whether the videotelephone could provide the necessary contact. The project involved a collaborative effort between two nursing colleges, local industry and Norwegian Telecom. Videotelephone equipment was placed at three different locations: two nursing colleges and one nursing home.

In addition to supervising students while practising at the remote nursing home, the videotelephones were used for sharing teacher resources at the two colleges. A lecture given at one of the colleges could thus be followed by students at both colleges. The videotelephone technology, originally developed for person-to-person communication, was used to teach a total of 60–70 students at two separate locations at the same time.

In order for this to function, some technical arrangements had to be made. Large 46-inch (120 cm) video monitors were placed at the front of the auditoria so that all the students were able to see the lecturer. The lecturers used computer prepared overheads which were displayed on a separate screen. For the lecturer to be able to see the students at both locations simultaneously, and at the same time operate the computer, a kind of teacher's desk was built (Figure 7.3). The 'desk' incorporated both the codec, the computer, a microphone, a video monitor to show the remote students and space for written notes and other kinds of prepared material.

Figure 7.3   *Teacher's desk – containing videotelephone, computer and microphones*

This desk also made operating the technology more user-friendly. All the cables could be hidden and the whole desk could be locked up and transported to another location.

With as many as 60 to 70 students present, interaction is necessarily limited. In order to create some feeling of togetherness, camera focus was shifted from time to time to the students instead of the lecturer. This was an important strategy to try and make it easier for the students to participate in a dialogue.

Another important strategy is to give the students easy access to microphones. Having to speak into a microphone when addressing the teacher is in itself frightening to many students. In this experiment there was only one microphone which had to be shared by all the students. It was placed at the front of the auditorium and any student who wished to speak had to leave his/her seat. This was a severe restriction on communication. It is therefore vital that microphones are placed as close to the students as possible, and in sufficient numbers.

Adult training in regional parts of the country

In one of Norway´s largest timber districts, Trysil, the local secondary school aims to become a competence centre for different target groups in the district. In 1989 a videotelephone was installed at the school, enabling it to offer distance education courses to adults who would otherwise have had difficulty in attending any kind of further education. Instead of having to travel to a remote school, these adults can come to the local school in the afternoons. There, qualified teachers from other parts of the country appear on screen and communicate with them.

In Trysil a special effort was made to give the students easy access to the microphones by installing one microphone for every two persons. The microphones were of the 'push to talk' type.

By placing the microphones immediately in front of the learners, the physical barrier of addressing the teacher was made as low as possible. An attempt was also made to minimise the psychological barrier by allowing the students to meet their teacher in person before the course started. This has proven to be of vital importance for achieving a good dialogue in all kinds of distance education, and it is highly recommended, irrespective of the type of technology being used.

In the various courses delivered to Trysil the distant teachers communicated with a group of learners, the size of the group varying from five to twelve persons. Due to the reduced quality of the coded and decoded picture, it was impossible for the teachers to recognise individual faces on the screen. The teachers felt the need to be able to see the whole group in order to have a feeling of reaching them all and to know that they were all actually present. At the same time they expressed a need for recognising the individual students with whom they were communicating. To fulfil this dual wish, extra cameras were installed and connected to the microphones (shared by two people) so that every time a student pressed the microphone button, the camera focusing on him/her would be activated. As soon as the button was released, the camera showing the whole group would be activated. This solution proved to be much more satisfactory to the teachers.

In this project, as in the TOVE project above, a 46-inch screen was erected to enable the students to see the teacher and the presented material in an enlarged format. The screen used in this project was also able to display a small picture within the larger picture. This gave the students the opportunity to see their teacher in a small video window while the rest of the screen displayed a written presentation made on the computer. The teachers in this project were all operating from a kind of desk as shown in Figure 7.4.

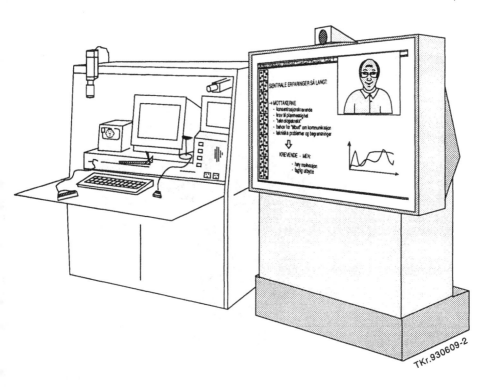

Figure 7.4   *Teacher's desk – version with a videotelephone, monitor for 'self-view', and computer; with large screen display attached*

As with the desk used in the TOVE-project, this was an integrated solution, where all the cables were hidden and all the equipment was easily operated with the aid of a few buttons. The desk contained a camera to film the person sitting in front of it, a camera to film documents or illustrations, a microphone, light sources, a monitor showing the person(s) at the other location, a monitor for 'self-view' and a computer. This desk could also be locked and transported and was specially designed for situations where the teacher is isolated from his/her distant students.

Integration of video, sound and data

Experiences from our field trials early on had demonstrated the need for simultaneous computer communication alongside video and audio communication in certain subject areas. This need gave the impetus to the development of 'Teleproff C200' (Figure 7.5).

TKr.930609-3

Figure 7.5   *Teleproff C200 – a workstation which integrates video, sound and data*

Teleproff C200 is a workstation which integrates video, sound and data communication. It has been developed by the Norwegian company NovaKom at the request of the NTR and is being used for distance education delivered to Trysil in the autumn of 1993. Both outgoing and incoming video images are displayed on the screen of a computer, and all external devices are operated by mouse and on-screen radio-buttons. To make operation of the equipment as simple as possible, a short menu has been designed with only a few functions to choose from. The applications are Windows-based and easily operated by a mouse.

Teachers of computer programming had at an earlier stage expressed a need for communicating on an individual basis with their students. To make this possible, the teacher has been given a control panel of the class on his/her screen. From this control panel the teacher can connect each student's microphone and at the same time activate his/her mouse and keyboard, thus enabling person-to-person communication between them.

It is important that operation of the technology is made as simple as possible. Communicating at a distance, through media, represents a new experience to most teachers. In addition to the technology, there are several factors in a such a situation that may place a strain on the teacher. It is important that teachers be allowed to concentrate on teaching and establishing a dialogue. In order that the teacher avoids spending too much time on operating the media during the lessons, it is important that he/she is given sufficient time to get to know the media. It should also be remembered that a situation of this kind is new to the students as well. Therefore steps should be taken so that the students are also allowed enough time to become acquainted with the media.

Collaboration across borders

Nordic universities are relatively small, between 5000 and 30,000 students. The departments are also rather small, and there are only a few doctoral students at each department. The local senior staff does not necessarily possess their main expertise within the subjects chosen by the doctoral students for their thesis. Thus, both staff and doctorate students would benefit from collaboration with other universities.

The Norwegian University of Technology in Trondheim and Chalmers University of Technology in Gothenburg (Sweden) have collaborated on a professional level for many years. With the introduction of ISDN and videotelephony a new possibility for sharing educational resources appeared. In the summer of 1992 videotelephones were installed at both universities. Courses given at one university could then be followed by students at both locations, and doctoral students receive guidance from experts at the remote university.

Via ISDN, videotelephony will be an easily accessible service, and even internationally the traffic costs will be relatively low. Cross-national collaborations using videotelephony, like the one established between Norway and Sweden, are believed to be of great educational value in the future, especially to students at small institutions with limited resources.

The pedagogical challenge

Two-way visual communication brings about a situation akin to ordinary classroom teaching. Many teachers therefore feel comfortable using this technology because it allows them to act in much the same way as they do in the classroom. It should be recognised, however, that despite the resemblance, distance education, no matter what technology is being used, is a totally different method for both teaching and learning. This cannot be stressed strongly enough.

Even if the teacher and students can see each other, they can usually only see each other's head and shoulders. In addition, they can only see a very limited part of the surroundings. The possibility for communicating with the aid of body language is therefore also limited. Teachers who gesticulate and move about a lot may find these restrictions difficult to accommodate.

A most important consideration for distant educators is the necessity of preparing the illustrations, pictures, diagrams and texts to be used specifically for the chosen medium and the current situation. Many teachers seem to ignore this fact. The camera used to film documents does allow the presentation of pictures and diagrams, but the presentation on a screen cannot be expected to equal that of a presentation on an overhead projector. This is particularly true of a picture that has been coded and decoded before it is presented on a screen. It is therefore crucial that the presentations are prepared with care.

Teaching with the aid of this kind of technology, therefore, requires careful preparations. Preparation is also vital for the pedagogical character of the presentations, particularly for the communication between teacher and students. Both the teacher and students need therefore to prepare themselves more thoroughly for the distant lectures than for ordinary classroom education.

Teaching at a distance through videotelephone technology, represents a new experience for most teachers. It is important that teacher-training colleges acknowledge the responsibility for educating teachers in the use of this new technology. Exploring all the possibilities the videotelephone presents in distance education will take some time. Particularly demanding is the task of developing new methods for presenting learning material through this medium, exploiting the possibilities for presenting visualisations and illustrations, and finding other media that may be used in fruitful combinations with the videotelephone. This is first and foremost a task for pedagogues, but it is important that technicians, organisers and educators co-operate in this matter. Hopefully, the experience gained in the Norwegian experiments will prove useful to others who are planning distant education projects with the aid of videotelephone technology.

Chapter 8

ISDN videoconferencing for education and training

James Lange

The benefits of videoconferencing

Videoconferencing allows a two-way sight-and-sound connection between two or more rooms, regardless of where they are located around the world. For meetings, the connection is literally the next best thing to being there; for education and training it is often better. For example, students in a videoconference teaching network in Western Australia said they could see demonstrations better in the videoconference than they could in face-to-face classrooms. Students in University of Hawaii videoconference courses did better on exams than students in face-to-face classrooms.

Falling equipment prices and the advent of ISDN circuits with call charges less than a call from a car phone, mean that the technology is affordable to most companies and educational institutions, even those of limited budget. They are affordable because these organisations can show a pay-back in productivity gains, reduced costs and increased revenue. They amortise their equipment costs in less than a year and begin to show real bottom-line gains. Before we examine the cost-benefit analysis in detail, we will look at the conceptual benefits.

First, imagine the effect of letting top-quality instructors teach in two or more classrooms at once, anywhere in the world. Students can have access to the best instruction available, regardless of where the instructors are located. Companies can import leading-edge thinkers into their offices and factories to stimulate their workforce. Universities can bring the top experts into their classrooms for a 15-minute guest lecture, followed by a question and answer session.

Even with the same instructors that are used now, educational institutions and training departments can make a dramatic improvement in the quality, the amount and the reach of instruction for a minimal increase in delivery cost. The saving of instructor travel at IBM's ISEN (Instructional Satellite Education Network) resulted in a threefold increase in instructor productivity. The Knowledge Network of the West (KNOW) reaches students in British Columbia communities who have no other access to tertiary education. Savings in student time at United Technologies Corporation (UTC) outweigh the added delivery costs by a factor of five or more. Even better, 40% more UTC employees enrolled in advanced courses, including some from sites where no-one had previously enrolled.

Videoconferencing operating costs

Videoconferencing equipment comes in a wide range or price-performance points. At the high end is satellite broadcasting, which requires a fully-equipped television studio (say £500,000 with a large earth-station, say £250,000) broadcasting via a whole satellite transponder (£1,000,000+ per year), with quality as good as anything you have ever seen on your television set. Such facilities can be rented for a few thousand pounds per hour. For interactive sessions, one either needs a similarly equipped studio, earth station and satellite transponder at the other end or, more often, one settles for a telephone return link for interaction. This is called one-way video, two way audioconferencing: the presenters can hear their audience, but not see them.

Whether rented or purchased, satellite seems, at first glance, beyond the means of many companies and educational institutions. However, it is widely used in education and training, and those who use it have a sound economic reason for paying the price: it is a more cost-effective way of teaching and training than sending instructors and students flying or driving around the country or the world. (For example, IBM's ISEN and British Columbia's KNOW use satellite broadcast videoconferencing.) Since they understand the cost-benefit economics of satellite broadcast videoconferencing, most companies and educational institutions who use it are also using the less expensive two-way ISDN videoconferencing. If they are not, they should be...and they should also be asking some hard questions of their media managers, their training managers and their accountants.

The recent growth of digital telephone lines has brought affordable two-way videoconference tools to education and training. In the late 1970s, videoconference equipment became obtainable at under £250,000. This equipment compressed the television-quality video (140 Megabits per second or 140 Mbit/s) to 6 Mbit/s then to 2 Mbit/s. Granted, the video quality at these lower speeds was not as good as regular television (which runs at about 140 Mbit/s) but it is as good as the early equipment was at 2 Mbit/s. The best analogy is to think of the difference between a good stereo radio broadcast (2 x 15 kilohertz) and the telephone (3.4 Kilohertz). The telephone is perfectly adequate for most business conversations, but we would not want to listen to an opera or rock concert over it. Similarly, the quality of video when it is compressed (providing there is a good compression algorithm and good pre- and post-processing of the video and audio signals) is good enough for any business meeting and most training, but we would not want to watch a ballet or a tennis match over it.

For training and education, compressed video is more than adequate for 98% of all applications. The exceptions are those where we want to show a very rapidly moving object – perhaps a boxer, a drill or a high-speed pump. There are also situations where one-way video is more cost-effective than two-way interaction, such as when their are a large number of receiving sites to be reached at once. But for the majority of training and teaching, leading edge companies and educational institutions (IBM, Ford, Stanford University) which had been using one-way satellite broadcasting began moving many of their courses to two-way, fully interactive compression equipment that ran on digital circuits.

Digital telephone circuits at 2 Mbit/s, though cheaper than satellite broadcasting, still have high operating costs. Today's leases are measured in tens of thousands of pounds per year and occasional use at over £500 per hour in Europe and over $1,000 per hour overseas.

In the 1980s, the arrival of AT&T's Accunet Switched-56 Service (with circuits in multiples of 56 kilobits per second or 56 kbit/s) in the United States meant that the operating costs dropped to just twice a normal telephone call. At the same time, PictureTel Incorporated (a company founded in 1984 by two MIT graduate students who had been studying video compression theory) produced equipment that could be rolled from room-to-room at under £100,000. The video quality at 112 kbit/s was as good as the early 2 Mbit/s systems. These simultaneous developments meant a very affordable two-way tele-teaching tool was available. Other manufacturers soon followed this lead, and the results are systems which can be purchased for under £50,000 and still deliver quality sufficient for education and training at the cost of two telephone calls or one mobile phone call.

ISDN (Integrated Services Digital Network) began to provide low speed, low priced, dial-up network on a worldwide basis at the same time that videoconference equipment price began dropping and quality improving. The progress was initially slow, as each country's telephone company had to agree on standard interoperating conventions and arrange bilateral commercial agreements, but ISDN is now a available in all of Western Europe, North America and industrialised Australasia, with new countries in Africa, Asia, the Middle East and South America joining every month. There are still a few anomalies: ISDN is based on 64 kbit/s circuits but Japan and the Americas still primarily use 56 kbit/s switched digital; German ISDN is different from English ISDN which is different from French ISDN, but Euro-ISDN is on the way and there are rate-adaption techniques that let Europe and Asia talk to Japan and the Americas.

The end result is that, using ISDN videoconferencing, a company or educational institution can make a two-way video call to anywhere in the UK for under £1 per minute, anywhere in Europe for under £3 per minute and nearly anywhere in the world for under £5 per minute. This proliferation in digital circuits that do not need to be leased, but can be dialled like a telephone and paid for according to the capacity and minutes actually used, meant a boom for videoconferencing. Since 1988, PictureTel Inc. doubled every year and other manufacturers also began to show spectacular growth. Of the over twenty thousand low-speed, dial-up systems installed around the world, over 80% have been used for teaching and training at one time or another. Although they have not been bought primarily for training, it is a rare training manager or University lecturer who doesn't recognise the potential of the equipment once it is installed.

Operating benefits

If ISDN videoconferencing can do so much for education, with so little cost and such a significant payback, why is it not headline news? Why are videoconference manufacturers just beginning to cater to Tele-Training and Tele-Education needs and why are they underwhelmed by the response from educators and trainers, even though there is growing demand from leading edge companies for better education and training? The reason for the slow response is a basic rule of all human endeavour: technology can not make anything happen – only people make things happen. The barriers to Tele-learning, as to any change in education and training, are the perennial three: ignorance, fear, and apathy or lack of encouragement.

As few of the current educators, trainers or company managers were exposed to videoconferencing in their own training, their ignorance of the technology and its benefits is both understandable and natural. It is the job of technology vendors (such as the author), consultants and educational technologists to educate these crucial target groups. After all, we can hardly expect someone to implement something that they do not understand and are slightly afraid of. Fear arises from ignorance. People are afraid to try new things because they are unsure of the outcome. They ask questions such as: Will the results be worth the effort, or will I have wasted my time? In an ever-leaner business and educational environment the cost of wasted effort and mistakes is high, particularly for individuals. It is safer to do things in the old tried-and-true ways. Doing things in a new way is risky. Apathy is partly a function of how busy one is already.

As companies and educational institutions have 'down-sized' and raised productivity standards, many are running so hard to just stay in place that there is no time to even consider the idea of trying something new.

ISDN videoconferencing is not the first technology to be billed as one that would 'revolutionise the way people learn.' Edison thought that the film camera would do that. Next came radio broadcasting, television, computer-aided instruction and interactive video discs. All of these were going to be the answer to all pedagogical dreams and all seem to have delivered less than they promised. Is videoconferencing any different? Well, yes and no.

Yes, videoconferencing is different than other educational technologies in that it does not require the instructors to change their old methods of instruction. That is, the standard lecture works just as well via videoconference as it does in person, sometimes better. A videoconference unit can be added to a regular face-to-face classroom and the instructor's productivity doubled, trebled, quadrupled or more without changing the teaching style. That has not been true of any previous educational technology, and most have foundered on this factor alone.

No, videoconferencing is not different, in that the promise is probably greater than the result. When we at PictureTel say that "our mission is to change the way the world learns"™, we don't expect that all classes will be taught via videoconference by the year 2000. We will be satisfied to equal the limited success that educational film and videotape, educational broadcasting, computer-aided instruction and interactive video disc have in education and industry.

Our chances of reaching these limited goals are excellent, for two simple reasons. First, industry has recognised that training and re-training, 'skilling the workforce' in American jargon, is one of the most important tasks they face. As they reduce the size of the workforce, companies are demanding that fewer people do the same large tasks by 'working smarter'. As product life-cycles grow shorter, the tasks and skills required of workers change more rapidly so that retraining is needed more often in a single career. All this is happening at a time when the standard of basic training of school leavers is falling. Hence, we have the most innovative and successful companies on our side, committed to training and re-training their workforce.

Second, government funding policies have required educational institutions to be more entrepreneurial. Universities, technical colleges and training companies are looking for a competitive edge that ensures that they win the contract, not their competitor. It is not just pride, it is economic health and possibly even survival. This is a powerful motivator to overcome the apathy that militates against change. Excellence of instruction is one way an educational institution can gain or maintain that competitive edge; videoconference technology – tele-teaching – is another.

Many companies contract for education or training from the institution that is nearest to them, even though they are aware that they might find better quality at another. Being close may mean that an institution is more responsive, but the primary reason for favouring proximity is the cost in student time. Sending an employee off to a campus for a day, a week, a month or a year is enormously costly in terms of lost productivity. Managers quickly see the advantage of having instruction on-site; the employee is back at the workplace five minutes after the class is over. Furthermore, if students have to commute a long distance for training, even on their own time, they become tired and their productivity is impaired.

One compromise is an in-house training staff, except that they are inevitably required to be jack-of-all-trades. A compromise of quality for efficiency. A better solution is to bring an outside expert onto company premises to teach the class. This may improve quality for the receiving company, but it reduces productivity for the providing institution. While on-site at Plant A, the instructor cannot also be available to students at Plant B or at the home campus. Nor is an institution likely to send their very best instructors out on such on-site jobs, unless the price is right. Indeed, a Nobel Prize winner is unlikely to go to Plant A for several days at any price.

Tele-learning, teaching via ISDN videoconferencing, solves these problems without requiring a compromise in quality or productivity by either the training institution or the receiving company. Institutions that are developing Tele-learning partnerships are building their competitive edge, and PictureTel is committed to helping by fostering the relationships between educators and industry. We have designed special equipment and a how-to guide for instructors to make the process easy. PictureTel Learning Partnerships start with the answers to just a few questions, with which we can get a company or an institution started:

Educational Institution	*Company*
1 In which courses are you best in class' (top 5 in UK)	1 What courses are most important to your success?
2 To which companies do you offer these courses?	2 Who teaches these courses for you?
3 To which would you like to?	3 Who would you like to have teach them?

Just answering these simple questions has already resulted in mutually beneficial partnerships between industry and educators. A few examples are cited below. Many of these institutions offer courses via videoconference that are included in the PictureTel Telecourse Catalogue, which means they are open for enrolment to anyone with suitable equipment.

Boston University (BU) works with United Technology Corporation (UTC) to give UTC employees equal access to a post-graduate engineering course, regardless of the students' location. UTC saves the man-hours that were once lost to student travel time. Participation gives BU a 40% increase in enrolment and hence boosts faculty productivity. BU paid for the equipment in six months and, as the UTC students are simply added to existing on-campus classes, BU shows a gross margin of over 50% on the tuition fees from UTC students.

Columbia University in New York City delivers post-graduate courses in electrical engineering, materials science, computer science and mechanical engineering to engineering professionals via PictureTel ISDN videoconferencing. A recent addition to their Master's Degree programme is Intel Corporation in Princeton, New Jersey.

Cornell University links their main campus in Ithaca New York to their branch campus in Manhattan to conduct simultaneous classes and to bring in guest lecturers via PictureTel videoconference. They also conduct student recruitment interviews over the system.

Massachusetts Corporation for Educational Telecommunication is using PictureTel videoconferencing to connect Bank Street College, the Educational Development Corporation, Integrated Systems Inc, Multimedia Research Inc, the State University of New York, the Southeast Educational Region Consortium and the Talcott Mountain Science Centre in a nation-wide project to improve science, mathematics and literacy training.

Metropolitan Life delivers executive seminars to five key headquarter sites, trains controllers and trains information service workers on new network applications via PictureTel videoconferencing.

The Swedish Adult Education Authority delivers courses to remote communities and re-trains civil servants on contract with Latvia, Lithuania and Estonia via PictureTel videoconferencing.

Stanford University delivers over 250 engineering courses to over 200 corporations across the United States, reaching over 5000 students per year. Recent additions to the two-way video course network include Sun Microsystems and United Technologies Corporation.

Other PictureTel Learning customers include: Arizona State University, Blue Cross/Blue Shield, Chevrolet, Caterpillar Commonwealth Edison, the Electrical Power Research Institute, University of Florida, Northwestern University Pennsylvania State University and many others.

Tele-learning equipment

There is a wide range of price points in dial-up (ISDN) videoconferencing. The main differentiators are the video and audio compression algorithms, the bandwidth accessible or required and the feature set available on the equipment. Equipment divides roughly into performance-based systems and cost-based systems and, within these categories, into room systems, desktop units and videophones. Performance-based systems are designed to provide the maximum video and audio quality and the most flexible feature set, then priced at a level that allows the manufacturer to make a fair profit. Cost-based systems are designed to meet a given price point (say US$10,000, for example) and then provide as good a video quality, audio quality and feature set as will permit a fair profit. In short, the manufacturer reduces performance in order to meet the price point.

Performance-based room systems are required for most teaching and training applications, though the smaller units will also be discussed. Cost-based systems are ill-suited to education as their lower quality video, audio and feature sets limit instructor control, student comprehension and classroom interaction. Tele-teaching and Tele-Training equipment should have a special feature set specifically designed for instruction.

There are many things to look for in a room system for teaching and training. These include:

- Teaching feature set that allows the instructor to operate the most common functions with the touch of a single button: for example, see-me, see my graphic, see my whiteboard, see other classroom.
- Compression software with the ability to run both higher-quality proprietary algorithms and the H.320 standard algorithm.[1]

---

[1]Some systems run only the H.320 cosine-transform compression algorithm that is accepted by the Telecommunications Standards Sector of the International Telecommunications Union as the current videoconference standard.   As with any standard, H.320 has deliberately frozen technology at moment in time (1988) so that competing manufacturers can ensure inter-operability with each other.   Major manufacturers have continued to developed proprietary algorithms that enhance or extend the standard to provide better audio and video quality and better feature sets. For example, proprietary algorithms allow acceptable video and wideband audio (7 kiloherz) while running at 128kbps, while the H.320 standard is Ò acceptable at that speed only if narrowband audio (3.4 kiloherz, 16kbps) is used.   Proprietary algorithms also contain special features, such as far-end camera control, which are valuable in teaching, but which are not included in the standards. Please note that some vendors without high-quality proprietary algorithms may recommend running performance-based systems at 384Kbps, in order to provide acceptable video and audio for teaching. This requires both additional equipment and higher operating costs. All of the major vendors (including Picturetel) sell an interoperating version of the H320 standard algorithm, though with some manufacturers, older equipment cannot be converted to run the standard without a major hardware change.

- Full-duplex wideband audio of at least 7 kilohertz – so that voices, sounds and music have a natural ring – as opposed to narrowband 3.4 kilohertz audio, which gives fidelity and range no better than the telephone.
- Dynamic echo cancellation so that two or more sides can talk at once, even if they change the room acoustics by moving about, moving microphones or entering and leaving the room. Beware of audio systems with clipping and/or echo when two sites talk at once or those that must 'retrain' when the room acoustics change.
- Full-CIF Video (CIF = Common Intermediate Format) as opposed to the much lower-quality Quarter-CIF (or QCIF) which gives poor quality when 'blown up' to the larger monitors or video projectors that need to be used in a classroom.
- Far-end camera control that allows the instructor to look around the distant classroom and to get a close-up of whoever is speaking, as we would in a normal face-to-face meeting.
- Camera presets that allow participants to set particular shots (a wide-shot of all participants, close-ups of key individuals or positions) on a high-speed pan-tilt-zoom-autofocus camera so one can look quickly around the room at the far end.
- Multiple Audio and Video inputs (say four or more audio sources and six or more video sources) without needing separate audio or video mixer.
- Forward and backward compatibility: can the system you buy today communicate with all previous systems from the same manufacturer?
- Flexible options, including a range of monitor sizes, cameras, document cameras, slide and computer integrators, microphone coverage and other peripheral equipment that can be adjusted to fit the classroom or training situation.
- Multipoint capability to connect three or more sites. Ideally, a 'meet-me' bridge, similar to that used in audioconferences, that permits sites to dial in from any network (the bridge handles any plesiosynchronous adjustments needed). Modern 'meet-me' bridges allow up to 240 sites to be connected, though 10-site capability will suffice for most education and training applications.

At the bottom end of the cost continuum are video systems that can operate on the standard (analogue) telephone. There are analogue videotelephones on the market at less than £400, but the video and audio quality is too low, even for business meetings, and the equipment meets none of the criteria established above for educational delivery. Unfortunately, due to the limited carrying capacity of the analogue telephone line, this quality is unlikely to be adequate in the foreseeable future.

Slow-scan (or freeze-frame) video can send high-quality still video over a standard analogue telephone line, with a good quality black-and-white picture transmitted in just eight seconds and a colour one in 35. The operating cost is a single phone call and the equipment can be bought for less that £10,000 per end. Slow-scan has been used to add vision to audioconference teaching and training networks for many years (e.g. University of Wisconsin), and will continue to play a role where only analogue telephone lines are available.

Desktop videoconferencing systems are beginning to be available. Cost-based units (say £3000) will have QCIF video only, few features and will provide a motion picture that is only acceptable in a small window on a computer screen. Full-CIF video that is integrated into workstations and personal computers with screen-sharing and file-sharing software, will become a valuable working, teaching and training tool in computer science, CAD/CAM, accounting and other computer and screen-based disciplines. (Without these features, desktop units are little more than executive toys.) As LAN circuits become more robust, with latency at 300 milliseconds or less and able to guarantee delivery of data packets in the correct order, both transmission and bridging costs will fall. By 1994, videoconferencing on the desktop workstation will make the science-fiction idea of a school-terminal in every home technically feasible. However, when such a teaching system becomes socially acceptable is another question.

Cost-based room systems that will not meet the criteria listed above will soon become available for under £15,000. They will have low quality (e.g. QCIF video, narrowband audio, and/or standard compression algorithm only) and few of the features required for teaching. However ill-suited, cost-based room systems may be used by institutions and companies that compromise on quality to satisfy a tight budget, using the theory that something is better than nothing. As with a suite of furniture, a suit of clothes, an automobile or a machine tool, compromising on quality may prove to be a false economy in a competitive world.

Performance-based room systems that meet all of the criteria listed above, are available today and already in widespread use in training and teaching. Depending upon the monitor size, number of cameras, microphones, features, peripherals and capabilities, their cost ranges from £20,000 at the low end to over £80,000 for something fancy. Such systems are also suitable for business meetings, interviews and other non-teaching applications. As will be demonstrated shortly, these prices are small compared to the benefits and are quickly recouped by companies and educational institutions that use them to increase reach, revenue and productivity while reducing cost.

Cost-benefit analysis

As mentioned, ISDN videoconferencing can increase reach, revenue and productivity in education and training, while reducing the costs. For example, let us assume *Company* (with offices and/or factories in Belfast, Birmingham, Bristol, Edinburgh, London, Liverpool, and Newcastle) that wants have the best education and training for its managers, engineers, sales force and factory floor. Add *Business School*, with a first-rate MBA, *Leading-edge University*, and *Dynamic Training Company*, to our assumptions. What do they have to gain from Tele-learning? Put another way, in terms close to the heart of a purchasing officer, what is the payback and return on investment for the money spent on Tele-learning?

Company's training needs

Management

Only 10% of *Company's* management have MBAs or other management qualifications, though all of these are in the three top-performing divisions. The 'get your own, on your own' policy hasn't worked, producing only two MBAs in the last three years. But the barriers to any other policy are: five of the *Company* sites are an hour's drive or more from the nearest quality MBA and four locations have no MBA evening courses. A local solution means three different quality levels, and correspondence courses take twice as long.

Sales

The 25-strong sales force rebel at internal sales training as internal instructors have no selling experience. Since sales training refreshes morale and improves performance, *Company* has decided to get a three-day sales training course run by professional sales trainers. Pulling the sales force together for the meeting costs £7500 in food and lodging (£100 per person per day), £2500 in mileage, train tickets or air fares (£100/person) and £200,000 in missed sales opportunities, as each sales person is on a quota of about £1 million (£5000 per day) and all 25 will be out of the field for three days in training, one in travel and hangovers.

Engineering

Senior engineers at *Company* graduated over 10 years ago and are complaining about how hard it is not to get out of date. The board thinks this may be one reason why *Competition* came to market first with a product this year, causing £2 million lost revenue. The board is committed to providing regular high-quality scientific and technical education. The leading-edge specialty needed is available in only three sites and the two top UK experts are willing to give on-site lectures (for a fee) but are not available for the next six months.

Factory floor training

Company is constantly training and re-training its 2000-strong workforce as new processes and procedures are developed and product lines change. At least 10% are in some sort of training at any one time. Local technical training is good in three sites, acceptable in two others and poor in one. Management has noticed that productivity and quality control is 15% better in the three sites where local training is good, and difference in quality control is the equivalent of £175,00 per site. Two of the sites don't have enough staff to warrant an on-site instructor (an on-site instructor costs £50 per hour) which means *Company* staff have to travel to a local campus or to another *Company* site for instruction, which costs a minimum of two hours travel-time per student. *Company's* best internal trainer has a new baby at home and is complaining about the amount of travel. The other three internal trainers complain that repeating the same class three or four time in a row means less time to develop needed new training.

Benefits of tele-learning

How can tele-learning via ISDN videoconferencing help meet these needs and what are the benefits?

Factory floor training

Tele-learning ensures the highest quality training, whether from internal trainers or from a technical college, is available in all locations. The training is the same in all locations, and can even be used to build corporate culture and unity among factory workers who rarely meet each other. It reduces student and instructor travel, trains larger numbers per instructor and cuts the amount of time instructors spend in repetition. Thus, tele-learning not only cuts costs and improves quality, it also raises the productivity and morale of the internal instructors, reduces the risk of losing staff to traffic accidents and lowers environmental pollution, though some of these benefits are difficult to quantify.

There are benefits of ISDN videoconferencing which are easily measurable. If *Company* can eliminate the on-site training costs (say £50 per hour) in five sites for 100 hours of instruction, the savings are £50 x 5 x 100 = £25,000. Eliminating a 30-session (15-week) travelled-to, on-campus course saves £480 per student in travel time, assuming the average factory wage is £8 per hour and the average productive time lost is one hour there, one hour back (£8 x 30 x 2). If another 100 students no longer need to travel, the cost saving is (£8 x 2 x 30 x 100) or £48,000.

As mentioned at the beginning of this chapter, tele-learning gave IBM a threefold rise in instructor productivity. Being conservative, we anticipate productivity of *Company's* four training staff to double. These four cost *Company* £30,000 each, which means a productivity gain of £120,000. These estimates would be very conservative, without even mentioning savings in travel costs for the trainers. Similarly, factory workers paid £8 per hour are worth considerably more per hour to *Company*, otherwise *Company* would be long-since bankrupt. Thus, our recovery of 6000 hours previously lost to training travel is worth well over £60,000 as a conservative estimate. In short, whichever way the productivity gain is measured, it is worth at least twice the actual saved travel costs, and this is fairly common for videoconferencing benefits.

The £175,000-per-site difference in quality control can't be claimed as a benefit of ISDN videoconferencing unless we can demonstrate that tele-learning can lift the quality control ability of the other four sites. The difference may be due to factors other than training. However, if improved training *can* improve quality control in the four weak sites, the potential strategic benefit (£700,000) will be more than twice the cost savings and the productivity gains combined. Although we will not include this benefit in our analysis, the benefit value mix is normal in videoconferencing:

Strategic advantage = 2 x productivity gain = 2 x cost savings.

Though we will see the same potential in other *Company* sectors, the cost-benefit analysis will not include the strategic advantage component and will use extremely conservative estimates of productivity gains, in an attempt to reduce the amount of figure-quibbling. However, it may be worth noting the cost-benefit ratios are so extraordinary that halving the productivity benefits claimed still leaves a payback of less than 24 months and a return on investment of over 100%.

Engineering
Tele-learning via ISDN videoconferencing can bring the 'best-in-class' scientific and technical education into *Company's* laboratories at an affordable cost. This not only cuts travel time and hence increases engineering productivity, but it allows the teaching content to be tailored to real-world problems. One of the top experts can be available for a one-hour tele-lecture per week to *Company* engineers (for the same fee charged for on-site lectures, but with no travel expenses) and the other top expert is running an on-campus lecture series that *Company's* engineers can attend by video. As engineers cost *Company* £20 per hour (£40,000 per year), on average, tele-learning saves £27,000 in time lost to travel for instruction (£20 x 15 engineers x 30 sessions x 3 hours). It is difficult to quantify the fact that five engineers who were too far away to attend on-campus instruction are now able to attend.

In addition to travel savings from training, *Company* finds ISDN videoconferencing can save 15% of its Engineering Department travel budget (say £50,000) by conducting production meetings, quality control, planning sessions and safety committee meetings via videoconferencing, as well as reducing the amount of time needed to resolve design-to-manufacture problems. Like many videoconference users, *Company's* engineers are able to boost their overall productivity by 5% or £100,000, assuming an engineering budget of £2 million. Others who use videoconferencing to improve time-to-market, such as Ciba Geigy and Ford, find time-to-market improvements as great as 20%, beating the competition by several months. However, as the value of this strategic advantage can be difficult to quantify in advance, a time-to-market benefit is not included in the cost-benefit analysis.

Management

Tele-learning via ISDN videoconferencing can bring an MBA course into *Company's* offices, making it easier for managers to attend by cutting their travel time. The courses can be run in the early evening, say 4:30 to 7:30, thus reducing the productive time lost to *Company*, and tele-learning ensures uniformly high quality across all sites. Indeed, Company uses the Telecourse to reinforce the corporate culture and to build bonds between managers who rarely see each other. This benefit is not easily quantified, as the impact of management training and shared corporate culture is primarily anecdotal. Similarly, improved management productivity has a positive impact in terms of competitiveness, survival and growth planning, but these gains are difficult to quantify. Nonetheless, our *Company* finds, as do most videoconference users, that they can reduce their management travel budget by 10% – say £100,000.

Sales

Tele-learning gives the sales force access to one of the most dynamic, up-to-date sales trainers in the UK, whose training includes a live, interactive presentation from overseas from one of the world's top sales consultants and live feedback from three of *Company's* top customers. They provide live follow-up sessions after one week, one month, three months and six months. Management are able to participate in the relevant portions of the training without leaving their home office, and sales support staff are also able to attend key portions. Sales discover that they are able to use videoconferencing to talk to their existing customers, and increase customer contacts by 7%, which has nothing to do with the value of the training course, but which results in sales that more than make up for the four-days lost sales time spent in training. Incidentally, *Company* management decides that the motivational aspects of an in-person training session are important, so the training is run as a face-to-face meeting and there are no cost savings on this item. On the other hand, Customer Service finds its ability to see the customer and the problem, and bring the appropriate expert in to quickly solve it, improves their customer satisfaction rating by 8% and their customer retention by 2%.

Costs

Benefits			
	Management	MBA Programme	Unknown
		Saved travel costs (10% of travel budget)	£ 100000
	Engineering	£20 x 20 students x 90 hours	£ 27000
	Other saved travel	10% of travel budget	£ 50000
	5% productivity increase		£100000
	Factory floor	£50 x 6 sites x 100 hours on-site training	£ 30000
		£8 x 100 students x 60 travel hours	£ 72000
	Internal instructor	Double their productivity	£ 120000
	Sales force	Customer contact up 10%, retention up 2%	£ 250000
Total benefit per year			**£ 750000**
Costs			
		Capital	£400000
		Recurrent	£250000
Conclusions	Payback < 12 months	Return = 187%	

Table 8.1   *Cost-benefit summary*

For the purposes of the Cost-Benefit Case Analysis, let us assume that *Company* spends £400,000 on capital equipment (a performance-based room system in each location and a multipoint bridge). We will further assume that the systems are used 30 hours per week (counting all applications, not just training) with 80% of the traffic between *Company* sites, 10% within the UK, 5% to the USA and 5% elsewhere, to get an operating cost (telephone company charges, maintenance, administration, etc.) of roughly £250,000 per year. Table 8.1 on the previous page summarises the overall calculations.

Subtracting the annual operating cost (£250,000) from the benefit (£750,000) gives a net first-year benefit of £500,000. Set against a capital cost of £400,000, this means a payback period of less than 12 months.

Return on investment

A simplified return on investment is obtained by simply dividing the annual benefit by the annual expenditure. Depreciating the capital equipment over three years, and including the opportunity cost of the investment (say 12%), we get an annualised capital cost of £150,000 plus operating costs of £250,000 for a total annualised cost of £400,000. Hence the return is £750,000 divided by £400,000 or 187%, an amazingly healthy R.O.I.!

But this benefit is not for *Company* alone; there will be a similar result for the *Business School* that is clever enough to be able to offer the MBA via tele-learning. We assume that *Business School* spends £50,000 on capital (one system), that 10 *Company* managers enrol, and that there is a gross margin of £2,000 per student. (Remember that these students are simply added into existing face-to-face classes via video.) In addition, the system allows *Business School* to offer the same course to other companies already using videoconferencing, whether here or abroad, which adds another 15 students at £2,000 each. The operating cost is nil for this application, as the customers pay the call charges. Furthermore, Business School faculty use the system to offer specialised short courses to this new customer base, which show an annual profit of £50,000. See Table 8.2.

10 new MBA students from *Company*	£20000
15 new MBA students from other companies	£30000
Profit from new short courses	£50000
Return = 200% (£100,000 on £50,000 capital)	£100000

Table 8.2   *Return on investment – Business School*

Similarly, the *University* buys a £50,000 system to offer the advanced technical courses, using videoconferencing to add 20 *Company* engineers to its existing on-campus courses, at a gross margin of £1,500 per student. It also adds 20 new students from other companies who use videoconferencing, and makes a £30,000 profit from specialised short courses it begins to offer to Company and to its other video-equipped customers.

10 new MBA students from Company	£30000
15 new MBA students from other companies	£30000
Profit from new short courses	£30000
Return = 180% on £50,000 capital	£90000

Table 8.3   *Return on investment – University*

Similar arguments show how a *Dynamic Training Company*, whether in concentrating on Sales and Management training or technical training for the factory floor, gets a pay-back in less than 12 months and a return on investment of over 100%.

Conclusion

It is no wonder that the best and the brightest have already begun to use ISDN videoconferencing for tele-education and tele-training.

Chapter 9

A French experiment in distance learning by ISDN: 'Le Visiocentre de Formation'

Jean-Louis Lafon

Introduction

The Ecole Nouvelle d'Ingénieurs en Communication (ENIC) has developed distant learning equipment using ISDN, called the Visiocentre. The school has used the equipment for its own undergraduate training, and found that students and professors are enthusiastic about it. This equipment allows the school to increase its effectiveness both by being able to use busy external specialists for teaching and by bringing the school nearer to the students in their workplace. In 1994, the ENIC will open a complete undergraduate course, taught at a distance.

ENIC is a school created in 1990 by two big institutions, the Lille University of Sciences and Technologies (USTL) on the one hand, and on the other, the National Institute of Telecommunications (INT), which provides higher education in telecommunication for France Telecom. Based in Villeneuve d'Ascq near Lille in the north of France, ENIC had for the academic year 1992-93, 400 students preparing for the State Diploma of Engineering in Information and Communication Technologies.

The students are in two groups:

- an initial intake (200 students) of young students who have passed the baccalaureat and will study for five years to get the diploma in the classic French manner
- a promotional intake (200 students) recruited in the business world, who enter the school with a 'Bac + 2' diploma and professional experience of at least four years. According to their level and available time, these students attend either a 'blocked' course which leads them to the diploma in 16 months, or a 'sporadic' course which allows them to spread their studies over two or three years.

This latter intake consists of very motivated people who want not only to study in a structured way, but also to get promotion within their firm. These students, aged between 25 and 40, coming from all regions of

France, attend ENIC for about one and half years. During this time, these students not only have to make a significant mental effort to succeed at their studies, but also have to stay in contact with their family and friends, and maintain the social and cultural activities necessary for their own personal equilibrium. As Lille is not in the centre of France, these students have to travel between school and home or settle near the school in order to avoid it.

To provide its 3000 hours of teaching, ENIC employs 20 full-time lecturers and 100 part-time professors. Depending on the subject, specialists may be needed to offer the required quality of teaching, and persons with such competence may have difficulty harmonising their timetable with that of the school. ENIC benefits from its links with France Telecom's *Direction de l'Enseignement Supérieur*, by having regular access to professors from telecommunications schools in Paris, Brest and Rennes.

At its beginning, ENIC created a research laboratory called LASER (Laboratoire Appliqué aux Systèmes d'Enseignements Répartis) composed of seven engineers specialising in various fields. The main goal of LASER was to devise and develop a complete system for distant teaching using NUMERIS, the France Telecom ISDN system. Lecturers not living in Lille must be able to teach ENIC students, yet the interactivity necessary for good course progress must be maintained. Furthermore, sound and image of good quality must be provided for full understanding by the students. The result of this research is the system called Visiocentre, which is described in the next part of the chapter.

When the technical equipment was ready, ENIC began to pilot its use, analysing the quality of the equipment in terms of pedagogic and social issues. The third part of this chapter deals with the context of the experiments made in the school. The fourth part presents all the results and the benefits that students and professors drew from the Visiocentre.

This equipment now allows ENIC to think in terms of spreading its activity: besides the two traditional channels, the school intends to open in 1994 a third way of preparing for the diploma, the distant learning way. This new capacity is discussed in the fifth part of the chapter.

The Visiocentre equipment

First of all, the Visiocentre covers a 300 metre square area including four distinct rooms:

- the 'Cabine Professeur' (CP): professor's room,
- the 'Salle Interactive de Visioenseignement' (SIV): interactive video students' room,
- the 'Salle Vidéo Câblée' (SVC): cabled video room,
- the 'Salle d'Auto Formation' (SAF): self training room.

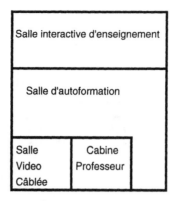

Figure 9.1 *Arrangement of Visiocentre*

This area can be used in the usual manner on 'local' mode: a professor can lecture to students (from 16 to 48 for the SIV, 16 for the SVC), because the two rooms are in the form of an amphitheatre with blackboard and stepped rows of seats. The SAF room, owing to its computer network, can also be used for sessions of directed or practical work with software, Computer Based Training and so on.

However, the Visiocentre area works primarily in 'distant' mode in the following manner:

- A professor wants to give a lecture to distant, widely spread students. These students are gathered in the nearest visiocentre SIV rooms. If in one of these Visiocentres, the number of students is larger than the SIV capacity, some students are placed in the SVC room working as an overflow room.
- The professor lectures from the 'CP' room. This room is connected to the SIV and SVC rooms by NUMERIS links, working at a rate of 256 kbit/s. The lecture starts: the professor speaks, shows, demonstrates; the students watch, listen, ask for further clarification, which the professor answers directly, going into more detail, and so on.

The professor's room

To give the lecture, the professor sits in a special room, a kind of office room in front of a special desk. A TV camera which is opposite, gives a 'French' movie picture of head and shoulders. In front on the desk, are:

- a touch screen computer
- a computer mouse
- a microphone
- a working area.

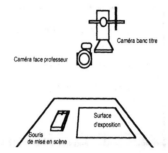

Figure 9.2   *Professor's room*

A bench TV camera is hung above to look down on the working area, and is centred on an A4-size sheet of paper in landscape format. The camera is provided with a zoom lens and positions are programmed to magnify documents up to A6 size.

The purpose of the computer mouse is to help view settings: with a click, the picture sent to the students changes from the professor's face to the view of the bench camera. A second click swings back to the view of the professor's face. The professor is thus able to diversify the picture for the remote students with a very simple movement.

The microphone in the CP is always active, which gives the professor authority during every moment of the lecture.

The working area allows the professor to use the bench camera to show every kind of document during the lecture: text, graphics, or technical material best seen with the zoom lens. But above all, the bench camera allows the professor the equivalent of a blackboard – a blank sheet of paper for writing formulae, text, or drawings. Of course there are some constraints due to the picture transmitting system: size of the letters, thickness of the pen, paper colour, ink colour. However, these constraints are minor and lecturers soon get into the habit of handling them. The desk has space enough to work progressively through the documents, with room for the used ones and the unseen ones.

The computer screen allows the professor to handle the view setting and to have control over the running of the lecture. The screen is divided into two main parts: the upper part of the screen is filled with icons symbolizing the different available audio and video sources in the professor's room; the lower part shows icons referring to the distant students placed in the various interactive rooms. The professor has only to touch the icons or to use the mouse in order to activate the interactive connection.

Various audio and video sources are available: videotape recorder, slide show, computer, video disc, laser disc, TV cable, phone and so on. Amongst these video sources, are the two cameras of the professor's room and the cameras of the interactive rooms. So at the professor's disposal are all the possibilities for making a 'televisual' lecture. To control which medium is required, the professor has just to look at the TV set on the left of the front camera. The picture on that TV set is what the students receive at the same moment. All the video sources are transmitted via a codec (Coding Decoding equipment) and the NUMERIS network (at a 256 kbit/s rate, even 128 kbit/s rate) to the SIV and SVC rooms.

Téléviseur   Téléviseur
de régie   de contrôle
terminale   des sites distants

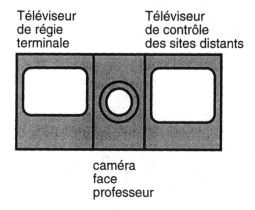

caméra
face
professeur

Figure 9.3   *Arrangement of TV systems*

The screen on the right gives the professor feedback from the remote sites, using the same NUMERIS link; each remote site is shown for 30 seconds, then it skips to another room and so on in a cyclical way. Thus the professor is able to have control over all the distant rooms – at least visual control!

The professor's room, in a soundproof office, is a true television office: the professor works there alone, without TV technician, separated from the students, even those in adjoining rooms who are actually very close.

The interactive room

As said before, the students are in a normal classroom; but during a lecture by Visiocentre, they are no longer looking at the blackboard but at a large TV screen in front of them. They see and listen to the professor whose face or documents and voice come to them by NUMERIS, handled throughout by the codec. While they pay attention to the lecture, a wide view camera is watching them, giving to the distant professor the image of the whole room.

One microphone is positioned on each desk between two students and has one button for signalling a question and one for cancelling the request.

(So, an interactive room can have from 8 to 16 microphones according to its size). If the student wants to ask the professor a question, he or she only has to push the call button and a green lamp will be lighted. A computer in the interactive room transmits the call via NUMERIS to the professor's computer, indicating with one icon the precise place of the calling student. That icon on the lower part of the screen turns from grey to green, giving a visual sign to attract the professor's attention. A gentle jingle sounds at the same time to give an auditory signal as well. The professor, now warned of the call for a question, remains free to consider whether to stop and take the question or to carry on with the lecture until an appropriate stopping point is reached.

When the moment comes to listen to the question, the professor presses, by finger or by mouse, the given icon which then turns red. At this time, in the calling student's room, the camera moves around its axis and zooms in on the calling student whose microphone then becomes active. The student's face and voice are transmitted to every participant in the other interactive rooms and of course, to the professor. During the ensuing dialogue, the professor may choose which camera to activate. Obviously this interaction mode is of the 'presidential type': the professor decides and controls the interaction. In further developments of the equipment, a four button box will be installed near each microphone: it will allow the professor to take polls in the class room (very clear, clear, medium, not clear). The professor's control screen will relay back immediately the answers dispatched from the various distant students.

The cabled video room

If it seems necessary to increase the number of listeners to the lecture in a given place, an extra 16 seat room is also equipped. In this room, a TV set faces the benches: the students follow the lecture by image and sound. But this room is not provided with camera and microphones that would allow interactivity. The students have only the benefit of the questions of their colleagues in the SIV. If necessary, they may send the professor a facsimile in order to submit their own questions. The professor can neither talk directly to, nor listen to these SVC students.

The self training room

In this room, 16 computers are connected in a network to a server, and four TV sets provide the sound and image from the professor's room. From here, the professor can send to the self training room server files, programs, exercises, or CBT, and can look at a student's file, have a better knowledge of the student's work, modify it, make comments on it. The professor can also enter into dialogue with the student by way of voice interaction.

Use of the Visiocentre by ENIC

The Visiocentre has been operational since October 1992: two centres have been set up in Lille and Lyon, cities which are 500 kilometres apart. This link was first used by ENIC for its own purpose and then spread to other educational centres.

ENIC has led two major experiments, one for its undergraduate course, the other during a training period for industry. The first one has been evaluated by a consultant both in terms of pedagogical and social issues.

For five weeks between October and November 1992, the school used the Visiocentre to start the courses for the students in the 'promotional intake.' These five weeks form only a part of the course – revision of basic mathematics, probability theory, physics, electronics, computer sciences. Of the 200 hours of teaching time during these five weeks, a third was carried out by distance lecturing.

The 75 students were divided between the two sites of Lille (60 students) and Lyon (15 students living mostly in the Rhône-Alpes region). For all the students, these lectures were the first 'at' ENIC.

About 15 different professors took part during these five weeks. Although their origin, age, and personal qualities were very different – younger or older, senior university professors or business experts – everyone took the time to become familiar with handling the Visiocentre through a short specific training, one month before the date of their lecture.

A multimedia service supported the professors by providing help in several ways: guiding them in the practice of view setting, creating lecture documents of different kinds for them (readable sheets of paper, video recordings, demo software, hypertexts etc.). Some of these documents were given to the students for self training use.

During the lectures, the professors used the different media to bring a new style into their repertoire. For instance, to illustrate probability laws, a professor played with dice under the camera's eye; the optics professor showed a diffraction experiment recorded in the laboratory; the mathematics professor used demonstration software to explain double integral calculus, and so on.

The training course lasted three weeks, once again between Lille and Lyon. The subject matter was telecommunication networks. About 40 students, coming from firms of the Nord Pas de Calais and Rhône Alpes regions, attended these lectures. The professors came principally from ENIC.

Why use the Visiocentre?

For ENIC, it was important to use the Visiocentre tool in a real situation: using it in these conditions gave the opportunity to check its qualities and faults, and its technical viability. From the pedagogical point of view, it was necessary to check that the equipment was well accepted by professors and students. By observing it in use, the LASER laboratory has gathered data to improve its ergonomics, already of a high standard, without letting it become too complex. A Visiocentre network is at present being built throughout France and even Europe, based on this feedback from the early uses.

Owing to this Visiocentre network, co-operation between educational institutions is increasing: every school can consider the possibility of widening its teaching resources by using qualified lecturers at a distance. By networking the professor's rooms, it is possible to give joint lectures (led by a principal professor and accompanied by expert reports). ENIC is working closely with the Telecommunications Schools of the France Telecom group: these schools will soon be provided with Visiocentre equipment, so that for the academic year 1993-94, the professors will not have to come from Paris or Brest to give a lecture at ENIC in Lille. This will mean time saved in travel and increased economy and efficiency.

These economic factors increase also when looking at the students, as the Visiocentre brings the school nearer to them. They no longer have long journeys, and can be more efficient because less tired. Furthermore, there are plans to bring the school into the company: the LASER development projects are to miniaturise the interactive room (SIV) in order to create a 'Visiosalle,' – a kind of desk equipped with a camera, a TV set, a codec, in front of a group of 6-8 persons with microphone and remote control. Thus the firm equipped with Visiosalle could follow lectures in undergraduate studies or training courses without its employees having to travel at all.

Evaluation of the Visiocentre experiment

Conditions for this first experiment were difficult, but the students were highly motivated. Both students and professors reported that the Visiocentre is very efficient, has a pleasant and simple ergonomics, allows good enough interactivity, and maintains the human aspects of the pedagogical act. During the experiment, tremendous solidarity developed between professors and students, and feedback from the five weeks was very favourable.

Difficult conditions

The experiment dealt with students entering ENIC for a course of undergraduate studies: they were employees who had to readjust themselves to situations of long and tiring training activities. These five weeks helped them to get into a structured and sustained working rhythm. The content of the lectures – mathematics and physics – was not insubstantial!

Efficiency

The Visiocentre is in some respects more efficient than the face-to-face lecture: seeing and hearing is as good, even better, and the Visiocentre formula leads to clear progress, no dozing, and no confusion. Contrary to the usual amphitheatre blackboard, the SIV monitor demands more sustained attention because no one can rely on the blackboard material remaining available very long before the professor wipes it out. Students reported that the lectures got better and better from one week to the next. When asked to compare their experience with face-to-face lectures, 61% of the students said they preferred the same lecture with the same professor by Visiocentre than in a normal classroom, although 84% said they would appreciate the professor using all video resources. In terms of image quality, students gave various scores, but always at least 6 out of 10. What 98% were in agreement about was that nobody falls asleep in a Visiocentre lecture! As for the ergonomics of the professors' room, most users expressed surprise at how easily they adapted to the altered lecturing conditions.

Interactivity

Everyone expressed the wish for more interactivity to make the system more human. Currently it is mostly a potentiality, used not very often, and not always easy to manage. Professors need first to be familiar with handling the different aspects of their lectures, in order to feel at ease with interruptions. When they master this new equipment, they intend to take time for animating the dialogue. Most of the students (84%) said they wanted to see who was asking a question. They think that makes the system more human and improves communication. However, they also wish the data they are discussing to remain on the screen.

About 70% of the students said they appreciated the times during the lecture when the professor initiated a dialogue, prompting the students to ask questions, make comments, or exchange points of view. However, they did not want it to take too much time. Some said that in mathematics and physics subjects, interactivity could be a waste of time. Nevertheless, from observation of the lectures, it was apparent that the interactive potential was used less than students and professors wished. Furthermore, everyone agreed that it was absolutely necessary to be available potentially. Interactivity is seen as vital for handling difficulties.

The students admitted that the interactive facilities were somewhat daunting: having to speak in public, with a camera searching for you, zooming in on your face and everyone else watching. Some were nervous that they might panic, splutter or forget the question they wanted to ask. For many students, therefore, the interactivity was very important – as long as it was someone else asking the questions!

Convenience

All the students from Lyon have made it very clear during the five week experiment that the great advantage they saw with the Visiocentre lecturing system was the reduction in travelling. Even more than saving time, they felt it saved travel fatigue.

> "It's rather better to arrive on Monday morning in the Lyon Visiocentre than in Lille on Sunday evening !"

Comparisons with face-to-face lectures

The Visiocentre emphasises what is already familiar (i.e. lectures), and at the same time, it leads to change. The atmosphere for the professor working in the small room is very different from that in a lecture hall. The atmosphere for the students in the Visiocentre classroom is also very different from that of the usual classroom. Firstly, the professor is prompted to create new documents and spends more time preparing the lecture. Secondly, the Visiocentre screen gives a different view from that in a typical lecture. The master's hand, seen in close-up view on the TV screen, generates reactions different from those when seeing the professor at the other end of a lecture hall. The professor's hand is useful to show on the screen: "you take this, you add that, it gives you the result you see here." This hand is the master's hand which gives life to the object by direct writing. This helps the student to connect with this object by identification with the master: the student sees the master's hand on TV and looking down copies on paper the same movement, the learning motion.

The rhythm of lectures is fairly similar, but it has a greater influence in the case of the Visiocentre. Documents shown in the Visiocentre are similar to transparencies and data shows in the classroom, but with a greater effect: in the SIV, when you look at the TV set, you see only this sheet. Consequently, mishandlings of the documents are more obvious and annoying, while at the same time, good use of material is more impressive in the Visiocentre. Similarly a long or boring lecture seems much longer and even more boring via the Visiocentre.

Certain changes in teaching methods have been noted through use of the Visiocentre: professors have the opportunity to try out new pedagogical methods and to create a new set of documents, graphics and other teaching aids. The most dramatic change is that the professor is no longer physically present in front of the students and much is altered even if both sides very quickly adapt to it.

Atmosphere

The atmosphere for the professor is very different from that of the usual classroom: the professor is cut off from students, staying in the intimacy of the transmission room.

> "You have many more resources; you sit down quietly; you can cut yourself off from the students and concentrate on the object of the lecture."

The Visiocentre strengthens the 'presidential' aspect of the lecture. Students must wait to ask a question and the professor can choose when to close the communication. However, although the atmosphere is conducive to concentration, the lack of feedback, the 'hum' of the normal classroom can be isolating for the professor.

For students in the Visiocentre classroom, there are at the same time two parallel realities: firstly, the lecture world with the professor speaking, and showing images, and students asking questions and being displayed on the monitor, and secondly, the real classroom world with students speaking amongst themselves while the professor is not able to hear them. This second world, parallel to the TV world, provides them extra energy for the intensive concentration of the 50 minute lecture.

Including visitors, the large classroom in Lille was usually full with sixty people seated one beside the other, looking at two TV monitors. In Lyon, the classroom was smaller, pleasantly furnished and the number of students was less than capacity. There was only one TV, well placed and clearly visible at the eye level. For these students it felt as if the professor was physically present.

Documents

Several kinds of documents with multiple and complementary functions were produced and used during the experimental five weeks: pre-prepared material which involved complicated formulas, illustrative slides which bring the subject to life, computer displays which added variety and increased redundancy, and partially pre-written material which was annotated by hand during the lecture.

Some of the professors sat in the classroom during a colleague's lecture to understand what was going on from the student's point of view. Others who had several sessions in the same week recorded their lectures and then reviewed them for ideas on how to make improvements. Generally, professors spent more time than usual preparing their lectures, to plan the timing and to write a 'screenplay' or synopsis.

The settings

During his Visiocentre session, the professor must play several roles almost at the same time: producer, stage manager and professor-actor. All the professors when interviewed spoke spontaneously about the absence of a blackboard. They also mentioned their compulsive need at first to fill any gap in speaking. If face-to-face lectures can be likened to the theatre, then Visiocentre lectures are more like the cinema.

From the students' point of view, the microphone settled on the desk in front does encourage interactivity: you can remain seated and simply speak in the microphone; you can first check with your neighbours whether the question is of general interest, and the physical absence of the professor is an advantage. Controlling this interactivity through changing the settings visible to everyone is, of course, formally the professor's job.

Lecturing rhythms

The professor manages the lecture rhythm and has control of the whole presidential system. After every 20 minutes roughly, a pause is made to give students time for asking questions, for summarising what has been done, and announcing what should follow. If two hours of lecturing is called for on the same day, the best schedule is a lecture at the beginning of the morning and the second one at the beginning of the afternoon. Two hour video lectures result in lack of concentration.

Further uses of the Visiocentre

To allow students to attend a distance learning undergraduate course in 1994, ENIC will put together tutored self training for individuals or small groups in the self-training room or in a distance mode. Each module consists of a pedagogical package of structured videotapes of 20 minute lecture sequences, printed matter such as booklets, exercises, reports and computer diskettes of CBT, simulators, and self tests. The student is always in contact with a tutor expert in the field. This contact will be either synchronous by Telemeeting or the Visiocentre, or asynchronous through exercise feedback, personal explanation by facsimile, answers to questions asked by other students, by Minitel file or program exchange via computer connection.

The ENIC course corresponds to 40 modules and will last about three years. In addition to home study, students can use the nearest Visiocentre and the in-company visiosalle, or self-training room.

Section 3

OTHER APPLICATIONS OF ISDN

The first three chapters in this section involve advanced ISDN audiographics systems used in a variety of education and training situations. Each of these systems has features and capabilities beyond the first generation of audiographics systems which required two telephone lines to transmit voice and graphics. These second generation systems integrate the facilities onto the PC and provide a common interface for ease of use.

Chapter 10 by Niki Davis is based on her work at Exeter University with Fujitsu's system (confusingly) called DeskTop Conferencing. One of her findings is that this medium supports the flexible learning patterns – both for students and in-service teacher training – which are now demanded. Another point she makes is that the shared screen facility of audiographics systems gives each side the feeling of ownership and control of the work. The synchronous nature of audiographics sessions is something that later phases of the Exeter project will seek to extend by an asynchronous messaging system.

This combination of synchronous and asynchronous facilities is one of the key features of the Co-Learn system described by Kaye in Chapter 11. Kaye also stresses the flexible, collaborative nature of the audiographics medium, as opposed to the teacher-centred, lecturing mode of some face-to-face and videoconferencing teaching. He goes on to review the history of the various group communications technologies which are integrated for the first time in the Co-Learn system – audio teleconferencing, audiographics systems and computer conferencing. A discussion of computer-supported collaborative work follows and the whole chapter is supplemented by an extensive bibliography of related research.

Chapter 12 describes the ECOLE learning environment which, like the previous two, emphasises flexibility of access as well as collaborative learning. ECOLE builds on the training advantages of CBT, but adds networked communications and multimedia tools. It is conceived primarily with 'one student per machine' in mind, and is hence a true desktop conferencing system. The applications of the system are Pan-

European with various training courses for telephone company employees, management and just-in-time training for maintenance staff. Some of these applications use the whole range of facilities available on the ECOLE system – synchronous videoconferencing and asynchronous messaging, as well as simulation games, self study and collaborative work.

The final chapter in the book, Chapter 13, describes the use of ISDN to extend the reach and functionality of a metropolitan area visual communications network. This sort of specialised use of ISDN demonstrates its versatility as an educational resource. In the case described by Beckwith, ISDN is one means of offering courses already 'networked' within London University, to distance students, particularly those in other European countries. ISDN, furthermore, offers a cheaper and more convenient connection to Asia and Australia than using satellite agents. Beckwith also uses ISDN to provide access to an image bank for uploading, accessing and browsing. It is interesting to compare the control features of the LIVE-NET visual communications network described (and designed) by Beckwith with other systems described in earlier chapters. In LIVE-NET, the audio-visual resources are controlled directly by students and teachers. Each site has full interconnection capabilities with up to four other sites simultaneously, without the need to control picture switching. However, this system is analogue and does not run over ISDN.

Chapter 10

ISDN technology in teaching

Niki Davis

Introduction

In many parts of the world there is increasing emphasis on school-based teacher education. Similarly there is an emphasis on children achieving their maximum potential. However it is difficult to locate sufficient expertise and resources in each school. One answer could be multimedia teaching from a distance, a form of computer supported co-operative work (CSCW) tailored to the needs of education and training.

New forms of more accessible multimedia communication have recently become available which permit new forms of working. In business and commerce, research and development are currently suggesting ways in which computers can support collaborative work between workers at the same location and those at a distance from each other. Gale (1992) notes that facilities of this new technology have been available for some time, but have not been taken up by users. Therefore the strengths and the weaknesses of this new medium require careful study within the social and curriculum contexts of education. The Exeter University Computer Conferencing Project was one of the first in the world to describe the use of multimedia communications over a distance to enhance education and training with ISDN-2. The project therefore fulfilled its aim to produce case studies in which this new form of communication and collaborative group work could benefit education and training. It also provides guidance for other uses of an integrated communications service.

The project

The project is based at Exeter University School of Education. It uses International Computers Limited (ICL) 386 personal computers with Fujitsu DeskTop Conferencing systems, linked with BT's Integrated Services Digital Network (ISDN-2) phone lines – see Figure 10.1. The systems are set up in three locations (the University and two schools) to enhance teaching and learning between university staff, student teachers, school students and school teachers. Business consultants have also participated in education-industry links through DeskTop Conferencing.

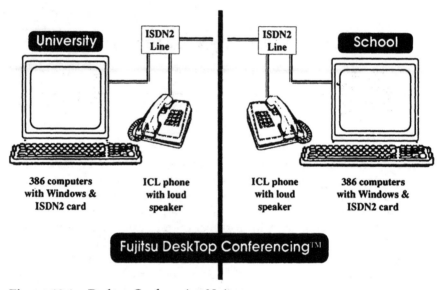

Figure 10.1   *Desktop Conferencing Unit*

DeskTop Conferencing (DTC) uses a pair of telephone lines to link participants at either end, sharing both voice and computer systems. Teacher and student can run the same software together, see the same screen, discuss it and even key in changes. In addition a flipchart facility is available and file transfer is extremely quick. To establish a link, one participant simply dials up the computer system and phone at the other end using the ISDN lines, which are part of the ordinary UK phone network. They then have a dedicated line for the duration of the call over which they can talk to each other and a second line which displays the graphics and text of one of the computer systems on the screens at both ends – see Figure 10.2. The call charge is the same as that incurred for the telephone in the UK for each phone line in use. The audio and data phone lines may be used separately or together.

The facilities similar to those provided through DTC have been available for some time using a range of software and hardware. However Fujitsu DTC contains the range of facilities in one computer work station with an associated loudspeaking telephone. It is important to have an integral package with its common windows interface and ease of use. Similar points have been made about point-to-point videoconferencing systems (see for example, Nordby, 1993).

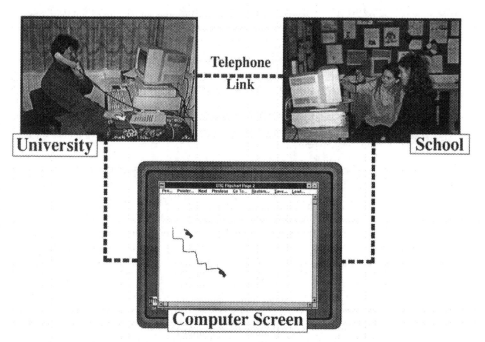

Figure 10.2   *A Desktop Conferencing link between school students and a student teacher*

The DeskTop Conferencing project has had three phases. The first gathered background information from the schools and student teachers, so that the case studies could be designed to fit within the existing work of the students and their level of skill. The research team also evaluated and became proficient in the use of DTC. This needs analysis would be important to the introduction of DTC in any institution (Davis, 1993a). The second phase piloted DTC and the final phase has consolidated the more useful applications, developed further ones and documented them.

Several activities occurred alongside one another. The general procedure was to provide a demonstration, to suggest activities and then to plan them in detail, encouraging staff and students to take the ownership. These sessions were monitored using diaries and discussions augmented by video and the documentation produced. A video stills camera was particularly powerful for catching critical incidents without extensive editing.

The rest of the chapter will provide a critical description of a range of case studies produced and the issues that they raise for teaching and learning. They cover a variety of applications: small group tutoring, one-to-one teaching and consultancy, anonymous or expert guest sessions, an Information Technology resources review centre and school-to-school information exchange. Subjects such as mathematics, business studies, design, modern languages and obviously Information Technology and communications were covered. In addition, the schools and university also received professional development. A future application, possibly of greater value to the university, is the supervision of students who are off campus to gain professional experience.

Teaching and learning through ISDN with DTC

This section describes the various ways in which the project developed the use of DTC and ISDN-2 to enhance teaching and learning for the university and the schools. The university provides both pre-service teacher education and in-service courses for practicing teachers. The project was designed to focus on pre-service teaching, but in practice also developed the in-service aspect in some depth. Such a move has been welcomed due to the changing nature of pre-service teacher education in the UK to increase the school-based component, such that by 1994 two thirds of professional studies will be located on school premises. The adoption of newer practices and support of teachers in schools has therefore become a priority for the university in its provision of high quality pre-service teacher education as well as in its support for schools. The following section describes case studies by application and by subject.

Small group tutoring

Group work is one aspect in education and training that staff aim to develop. When bringing a group together the first activity is frequently designed to 'break the ice' so that members will be able to establish working relationships. It was therefore felt that such an activity would also be important in DTC and it was merged with an introduction to the DTC software so that a range of facilities was introduced in a meaningful and enjoyable way. The 'ice breaker' case study is the result of this work and the riddles used in the exercise have proved educationally valuable. It was also possible to place the riddles into a variety of subjects and levels in the curriculum.

Science
A pair of student teachers prepared a spreadsheet model with which they tutored two school students. The model developed the students' knowledge of the Periodic Table, which they were studying as part of their course. The students' response developed the student teachers' knowledge of teaching. The student teachers also learned about the cross

curricula applications of Information Technology in the school curriculum. The students, keen users of IT, had the opportunity to develop their skills within a subject context, which they welcomed. The project thus responded to an aim of the school which was to address the criticism that students who were more able in terms of IT should be further developed and encouraged within their education. For the student teachers, DTC also reduced the time they would have had to spend commuting to the school. The inexperienced users of communications found it challenging and exciting.

Mathematics
Four of the year 8 students (12 or 13 years old) have been tutored in pairs and as a foursome on a variety of mathematical topics to extend and enhance their existing course. Symmetry and geometrical pattern have been explored using a light pen on the 'flipchart' shared between the university and the school alongside the audio phone line. These students have responded very favourably and the sessions have led to professional development for their teacher in these new approaches to teaching and learning. Older students (aged 16) have received some extension work to prepare them for a more integrated view of Mathematics appropriate to A level study. The graphical, numerical and algebraic work involved curve sketching on the flipchart and collaborative use of a spreadsheet with the University tutor. In the words of a student: "It is very good. Interesting, different and quite enjoyable."

Design
Four of the year 10 students (aged about 14) studying for GCSE Design and Realisation were introduced to Autosketch for Windows, which is a comprehensive, professional, technical drawing package. A teaching method rarely mentioned is that of successive refinement. Experts in different domains share their knowledge within the process of developing an artifact. An example between computer novices who are subject experts with their opposing partner (computer experts who are subject novices) has been described for spreadsheets by Nardi and Miller (1990). In this case, school students 'taught' Autosketch to a student teacher with expert knowledge of Design and Technology. The student teacher then helped the school students to refine their application of Autosketch using expert knowledge of the design process. The student teacher also learned about the way in which the school students understood both the software and the design process at various stages. The process continued to refine the artifact while also successively refining both the students' and the student teacher's skill and knowledge.

Business Studies
Four students participated in a similar venture with two student teachers. In this case the software was Aldus PageMaker used to create a logo for the students' mini business. The student teachers provided advice to the students on the design of the logo and the students' process of learning and skill in using the software were valuable to the student teachers.

Anonymous or expert guest sessions
The Business Studies students mentioned above also received advice via DeskTop Conferencing from a business person regarding their mini business. This authentic advice is sometimes given on visits, but DTC could make such outside expertise very accessible where it was also installed for business purposes.

Anonymous simulations
DTC was used along with other media for school students, aged 12–13 years old. They took part in a half day role play called 'Desert Storm' which was a media simulation set in Riyadh at the time of the Gulf War. It was organised by Postgraduate History student teachers under the guidance of their tutor. Many media were used as stimuli, especially the CD ROM 'Desert Storm'. Student teachers of history took roles such as the Emir and Schwarzkopf while others organised the day. Students became advisers to the principal characters or newspaper reporters with their varying points of view. The DeskTop Conferencing link provided an extremely stimulating 'Reuters' news with interviews from various places in the world, e.g. USA, Russia and UK. The anonymity of the medium was used to full effect by student teachers playing the role of operators with suitable accents and documents to transmit. Much was learned in three subject areas: history, media education and Information Technology and in the debriefing discussion a number of students found it surprising that they had not been communicating with Russia and New York! This case study is reported in detail in Wright *et al* (1993).

The benefits of DTC for teaching and learning

There must be a necessary level of IT literacy and ethos in a school to exploit new means of remote teaching through ISDN. It also appears to involve a particularly intense nature of teaching. On viewing the Exeter University Computer Conferencing Project, a consultant for the National Council of Educational Technology, Jon Coupland, commented:

"What is the particular, distinctive nature of the interaction between student and teacher when using ISDN? It may be something to do with having access to a teacher who can support but not interfere. This all links with the locus of control in the learning situation which is at the heart of much flexible learning."

It was also clear to the tutors involved that their 'links' with students were of an unusually personal and intense nature. It may be that students require less class contact time with staff using this medium, but we have no data on this at present. It could prove to be as important a cost saving for learners as it has for staff in-service training.

The students too were enthusiastic in their comments. For example:

"Brilliant"

"It was a more enjoyable way of learning and more effective."

"Anyone in secondary school could use it, but not in primary."

The unusual enthusiasm of the girls for new technology in the project was noted. They enjoyed using the technology and the fact that it wasn't just the keyboard – there was the light pen, the stills camera and video-disk. One teacher is now using these in her classroom having seen them used on the project, so work with learners also leads to staff and curriculum development. The teachers did not feel that the technology was a barrier to any of their students, even those who had never used computers or those of low ability:

"I don't think ability would be a problem with the hardware or speaking on the telephone. However the sixth form students did note that their collaboration with students in the other school made the importance of good communication very clear to them. This led them to some valuable reflection."

One to one teaching and consultancy

In the area of professional development in IT, several staff were tutored in the application of Aldus PageMaker. The support staff thus became able to produce higher quality, better designed materials for their schools. Senior staff received tutoring which enabled them to have an overview of the possibilities of Aldus PageMaker, still video camera and other products in order to improve the production and management of resources.

In-service training

Both school principals noted that in-service training was the most easily identifiable benefit that related to current budget practices. As one of the schools was proceeding down the independent (grant maintained) route, the principal felt that the availability of in-service and technical support through ISDN would be 'magic':

> "Just say at the end of a working day you plan a 3.45 pm encounter (link) where the maths tutor in the university and my maths staff here engage in a very intensive piece of in-service. Absolutely top rate."

The schools feel that the quality and relevance of in-service courses would be easy to identify and could be adjusted to their needs with great flexibility. Support available through DTC is easily accessible and where it is found to be of inadequate quality it would also be easy to discontinue.

The teachers also felt the in-service support was very valuable. A mathematics teacher explained the importance of multimedia when learning to use the still video camera via DTC link. She felt that it was better than face to face tuition:

> "I'm not very au fait with technology at all, but just having those pictures on the screen and Bruce's voice was marvellous. When someone is explaining how to use something standing beside you they often take it over from you. But using the link you have to cope with it. You have the verbal and the visual information and no one is taking it away from you. They try to explain to you instead. I did not find it a second best. I found it very useful."

The staff development for classroom assistants was also found to be very valuable. When asked to compare it with similar support from school staff the comment was "That wouldn't have been so inspirational." In future it may be possible for the school to share their expertise with other schools, but a continuing need for a hub, such as the university, to provide motivation and co-ordination is seen as essential by all.

Information technology resources

The Knowles Hill school principal, Colin Pope, saw the Computer Conferencing Project as a way of enhancing the school culture where the library is a resource centre to facilitate learning and supported study. He particularly welcomed the input of expertise from an agency outside the school. A similar description was given by Bill Ridley, principal of St Thomas' High school. He gave the following description to guide heads of other schools considering the purchase of DeskTop Conferencing and ISDN:

> "DeskTop Conferencing adds a further dimension to the school's resource centre, which is where students and teachers have access to resources which bear on their particular subject or problem. It can give them insights which would almost certainly not have been otherwise available. One of the things that DTC can do is to teach both the student and the teacher that there is more than one set of correct answers. Outside agencies are available through DTC, such as tutors, student teachers or business consultants."

DTC can provide access to software, databases and other resources at a distance. The software can be running on the distant computer and thus protect copyright. Personal mediation is also possible, as occurred in the professional development of the senior teacher. Here a view of Aldus PageMaker and other products in the hands of an expert persuaded the senior teacher of its place in education. Similar activities were possible for resources on CD-ROM with mediation through DTC. The senior teacher felt that such a review service could become extremely important given the complexity and expense of resources advertised today.

In addition technical mediation or hot line support is a powerful facility. Although remote fault finding and repair are common on large computer systems, such a facility is rarely offered to users of personal computers. DTC permits such a service with accompanying explanation of the problem and its remedy. One research assistant explained to the senior teacher in response to his concern that technical faults might develop with software:

> "I can take control of your computer from here and use it. It is really powerful and I have seen it demonstrated by ICL technical support to solve a real problem that I created by having an extra space in a command, which really messed up this computer. Justin (from ICL) took control of my keyboard in Exeter from (his location) in Windsor. I watched him (access the directories and put the file name) right and he explained what he was doing as he did it."

Brendon Cuddy, the senior teacher commented:

> "Technical support without any need to come out to the computer. That's excellent."

This senior teacher in Knowles Hill School was also impressed with the adaptability of the system:

"There is benefit in being led by the hand on occasion, trying out new software before purchasing it or linking several schools with each other. We are in a video age. It seems to be inspirational in itself and provide quality material. Students who are interested in the task generally perform better."

Issues relating to multimedia teaching at a distance

The use of multimedia communication can permit participants to glimpse new forms of education and training, where learners can collaborate with their teachers in a way which has been rare, even when sitting side by side. It is almost as if the voice link plus the window of the computer screen permit both to feel as if they have ownership and control of the work being produced. The following headings will be used to discuss the issues which have emerged: timing, navigation and control, information and co-operation.

Timing

The synchronisation of DTC requires care. The thread of communication is disrupted by delays over 10 seconds when participants wait for a response. The delays are reducing with each new version of the software, but developers using multimedia, especially video, should plan to adapt their style to the medium. The delay was disruptive because it was not expected. Users expect a delay with post as they have a mental model of a postman delivering the letter. The user does not expect to work on the reply immediately. In contrast a user seated at a workstation expects it to run at its normal speed even though communications are involved. Indeed, the computer screen was expected to synchronise with the immediate voice signal received and transmitted by the telephone.

Rather than explain this difference to the participants, the tutors responded by making the learner's machine the host or 'chair' and attempting to predict the outcome of their actions and to speed the interaction in this way to acceptable levels. Despite this, more experienced users noted that the lag of the computer screen made the sessions less interesting, although they remained useful. Later versions of DTC software will reduce delays further; nevertheless, a small time lag may remain inevitable.

The disruptive effect of the delay was reduced using a more detailed lesson plan than is common for tutoring. This benefited the student teachers, who became more aware of the phases of a lesson and the participation that they should expect from their students and had more time for the less obvious aspects of teaching. One of the student teachers commented:

"In fact it was much quicker than I had expected and I found we could use these 'waiting periods' to talk through what we were going to do next and continue to build up a relationship."

However the delays meant that the lesson took much longer than a face to face session.

At the other end of the time scale, scheduling the DTC sessions has required considerable care. Four systems were used across three institutions with several groups participating in each location. Each institution has its own timetable. In the pilot phase much of the work was done outside the normal timetables of all three institutions. Later it became part of the school timetables, although this disadvantaged the student teachers. Like many forms of IT in education, DTC requires flexibility of time and increased autonomy of learners (Cory, 1991). An asynchronous form of computer conferencing could be of value in solving this time-tabling problem between participants in different institutions. Using such a system, links and topics could be negotiated at different times to suit each participant. A later phase of the Exeter University Computer Conferencing Project is exploring complimentary applications of different forms of electronic communications.

In-service training, on the other hand, has benefited from the flexibility of time. It is relatively easy to find staff cover for 40 minute sessions, whereas a longer absence for courses away from the school premises. is much more problematic.

The intensive and tailor-made form of in-service training received by staff has had considerable benefits for the staff involved, although these are not easy to evaluate. Jackson notes some of these in his model 'calculating the benefit to the individual':

> One of the most important functions of training and development is to bring individual and organisational objectives together. (Jackson, 1989, p. 92)

It is clear that support staff improved attitudes to themselves, their tasks, to other people (staff and students), and to the institution. Their new skills have permitted them to assist staff and students in a more meaningful way and to relieve some of the work load of more highly paid staff. Senior staff have also benefited in this way although, as this is a higher level of skill, it becomes even more difficult to quantify. Senior staff are under enormous time pressure, which means that they are less likely to choose traditional in-service training taking them away from their work. For this reason, flexible in-service training may appeal especially to them. It may also reduce sick leave and turnover of staff.

Navigation and control

DTC requires several programs to run concurrently on both computer systems which are linked by the ISDN-2 telephone lines. Users therefore develop a complex mental model of the technology underlying DTC and on several occasions we have noted that such models were incomplete or faulty. For example, a user may be surprised to find after a DTC link has been disconnected that the learner's file that had been shown in the incoming screen is no longer available. Intellectually the user is aware that it is on the distant computer, but the immediacy and personal manipulation of the file make that hard to accept.

At times it was difficult to visualise with whom the link was being made and to locate various software tools. The project therefore assisted navigation by standardising the desk top arrangement across institutions while colouring each uniquely. In this way users could recognise and remember as a background cue to the information with which institution they were linked by the colour of the incoming screen. For example, Knowles Hill school used a pale pink desk top with red borders.

The development of DTC appears to have been influenced by the model of a chairperson chairing a committee, rather than the relationship between teacher and learner. In a classroom it should be immediately obvious that the teacher is in control, although the learners may be given as much autonomy as possible. DTC can be run in a mode in which the 'chair' has control and the 'participant' has to request to change items. The software even made the 'chair' confirm a new choice of pen colour by the participant. This is inconsistent with the required learning style, which should enhance the autonomy necessary for teaching at a distance. As noted above, tutors tended to put learners in the 'chair' so that they could show work they had developed on their computer with a greater speed of response. The need to request permission is also inappropriate for consultancy.

The question of control also raised longer term issues. The teachers felt that the ethos of the school was important. The head of resources in one school noted that if the pupils were 'sat down' in rows in their classroom there would not be any point in having an ISDN link. The school would have to have a certain amount of flexibility in its approach to teaching and use of the library. Most progressive secondary schools in the UK do have this approach. The students agreed that a well motivated learner is essential because, as one student noted:

> "You have to be more responsible. No one is forcing you to
> sit down. I could walk off but it is because I wanted to learn
> that I am here."

In fact, as the DTC system was placed in the school's resource area, the normal procedures for monitoring students in this area were in force.

Support staff would intervene if the student went blatantly off task. The importance of the student's statement is that he felt autonomous and was happy to use DTC with a tutor in the university.

Information

A video view of participants' faces is assumed by many to be the most important source of information between teachers and students. Although the students were curious, the lack of this video has not been disruptive because the focus of our tasks have been on the screen and input devices. The task is therefore a key to reducing the need for video. However, participants do lack cues which indicate what others are about to do next. In face to face situations such information is often provided by the direction of a glance, the movement of their hand or hesitation. This lack of information was noted in a tutor's diary after he had viewed video tapes of both ends of the DTC link:

> "Although Vi was doing most of the talking, Di was doing the deeper thinking. In fact Vi was quite often relaying, unbeknown to me, Di's answers. Assessment of group work would be very difficult through DTC. Di was quite often doing a significant amount of pointing at the screen, which of course I am unaware of. I think I should ask them to share the light pen and mouse more equally in future, otherwise Di will get very bored as Vi takes over. The students have become confident enough with the technology to instruct me in the use of the flipchart when I am not quick enough. (Vi told me that I would have to open a fresh page as I had the chair)."

The ICL phones with extra buttons, mike and loudspeaker work well, permitting groups to participate at each workstation. Occasionally listeners misinterpreted what they heard, for example, "Eh?" for "OK" which caused some confusion. This caused the tutor to repeat an explanation rather than move on to the next point. The technical explanation is that the loudspeaking phone switches to the phone with the loudest sound. Where the first syllable has less emphasis or overlaps with a noise from the other end, the conversation can be cut with unfortunate results, as above. In addition the direction of the speaker's mouth is frequently towards the screen or other people rather than the microphone, so speech may become indistinct. Similar problems have also been noted in a Norwegian study using video-phones with voice switching (von Tetzchner and Holand, 1993, p. 37). Tutors also noted that their teaching manner was adversely affected by having to speak to a microphone:

> "I was very surprised to see the effect of (my) using the hand
> set, rather than the telephone speaker, at both ends. It
> affected my posture (more relaxed), tone of voice (quieter),
> teaching style (more supportive and collaborative), and
> teacher-student relationship (more positive and informal)."

The clarity of speech is particularly important for situations where participants find it difficult or complex, such as when conversing in a foreign language or about a complex topic.

At the other end of the scale there are times where the teacher suffers from an information overload which is similar to an over-busy classroom, but also different. The tutor has several different channels of information to attend to: the local workstation running at least two different applications; the incoming window from the distant workstation; the telephone hand set's buttons and lights; the audio phone line which may have a number of voices on it; and the non computer based materials. Using this information the teacher will be building a mental model of the learners and using that to guide actions. Where significant dissonance builds up most tutors suffer an information overload. However, the project has tried to avoid this by structuring the sessions more closely than would be normal to our teaching style and by continually seeking confirmation from the distant participants that their workstation is behaving as expected. In contrast, younger participants rarely find this a problem. Today's video generation are happy to switch channels frequently on television or with computer games.

The anonymity of computer conferencing has been shown to produce more extreme behaviour (Spears, Lea and Lee, 1990). We have found little evidence of this possibly because the communication is almost immediate and the voice connection decreases anonymity. We have also used the effect to simulate events, such as the Desert Storm case reported earlier.

Co-operation

Group work skills are valued highly by employers. A group task was the basis of the ice breaker exercise right at the start. It was merged with an introduction to the DTC software so that the range of tools available (flipchart, software sharing and file transfer) were introduced in a meaningful and enjoyable way. The ice breaker activity was based on a "Who am I?". Clues provided in a variety of media: jumbled text, pictures and sounds, objects or people relevant to the students and subjects. Participants at either end of DTC worked in friendly competition and through this developed a working relationship with each other and the software.

Individual differences in using multimedia communications were indicated for gender and age. Girls seemed more ready to communicate than boys. They were also more likely to act upon a suggestion from others in their group and from distant groups. Littleton, Light, Messer and Joiner (1992) suggest that the task and the 'audience' have a large effect on computer-based work. With DTC, the ability to communicate is also a factor. One case study appeared to the student teacher to be unsuccessful: teaching French to low ability students whose poor French meant that they were too shy to participate effectively. The school students' reluctance to verbalise resulted in the student teacher receiving very little feedback and she therefore learned little from the session. Furthermore, she didn't enjoy the session. However the school teachers felt that the session was very valuable for those low ability students. Having the extra time and kudos of participating with an authentic French speaker in the university increased their awareness of French as a language and their motivation to use it. Their low level of communication was normal.

The teaching method of successive refinement, in which experts in different domains share their knowledge within the process of developing an artifact, was described during case studies for Design and Technology. It is a clear illustration of the changing role of the teacher in relation to flexible learning as facilitated by new technology. The student teacher and the school students are co-operating over the learning task.

Cost-benefit analysis

The Exeter University Computer Conferencing Project has provided the first case studies in the educational use of British Telecom's ISDN-2 service in the world. These case studies demonstrated that the ISDN-2 service can be used to good educational effect, but the question of cost effectiveness remained. An additional study was therefore commissioned by the Department of Employment Learning Technology Unit (Davis, 1993b).

The data for the report was gathered through inspection of documents gathered throughout the Computer Conferencing Project and interviews with staff in the university and schools involved, as well as with external consultants. British Telecom, Fujitsu and Ascotel staff and publications provided information on costs and technical details. The report provides evidence that ISDN-2 and DeskTop Conferencing is cost effective in education. Schools felt that the previous low use of electronic communications in the curriculum related to the lack of predictable costs or results. These were clearly predictable using DTC.

Many secondary schools and most universities have switchboards which may be upgraded to include ISDN lines in addition to the current analogue ones. Ascom provide switchboards which mix and match in this way. Using such a switchboard it becomes possible to switch and even duplicate information to a number of locations in the institution.

Distributed group work becomes possible for administration, teaching, learning and research. Because of their digital nature, it may become possible to link other networks, such as computer local area networks and the cable networks installed in a few cities.

The integrated nature of ISDN means that the same lines can be used for many different purposes by many people in an institution. In this way the cost of the system could be more easily justified, although its integrated nature means that it is difficult for administrators to gain an overview. Institutions with a senior management team who believe in flexible learning and applying new technology, will be the first to benefit from ISDN.

The report made the following recommendations:

1 ISDN and DeskTop Conferencing can be cost effective for secondary schools and universities, providing the institutions have an ethos which welcomes innovation with flexible learning and new technology. Such institutions should start to consider adopting these new communication technologies.
2 ISDN applications and the network itself is in its infancy. New users require more information on the applications that can be integrated. Funding should be provided to take potential applications further and so develop commercially viable ISDN services.
3 Potential users of this technology will find ISDN difficult to understand and to cost. Initiatives are required to increase understanding and participation, for example, sites which can demonstrate the equipment with real applications and information created specifically for different audiences.
4 Early users of the technology will continue to require support and co-ordination from an enthusiastic centre. However it will be difficult for the centre to become self funding in the short term.
5 Telecommunication providers should create educational rates suited to the needs of this market and also recognise that benefits will accrue from the increased skills which learners will thus bring to business and commerce.

Conclusions

DeskTop conference via ISDN-2 can provide a rich and varied resource for education and training. This new technology can perform a number of roles in addition to the current uses of a pair of phone lines and a powerful personal computer. It can extend the curriculum and be a vehicle for in-service teacher education, in addition to providing insights into the process of teaching and learning for student teachers and research. The value of the system depends on a number of factors: the motivation and application of the learner, the skill and art of the teacher or mediator, and the value of the resources available through the link. Communications do not easily become a permanent part of the educational curriculum (Davis, 1993c), but the possibilities stimulated by this project do suggest a new and valuable approach. DTC can provide extensive support for higher level skills, particularly those relating to software applications in the curriculum. Timetable problems change. Travel time and cost are certainly saved.

However, there are costs in terms of increased organisation, resources and access. The latter costs will vary with the institutions concerned. The need for teachers or resources will not be reduced, but multimedia teaching at a distance could extend and expand the curriculum and the quality of teaching and learning possible in secondary, further and higher education and training. A package of applications together will justify the expense. The drawbacks mentioned will not prevent DTC from becoming a valuable tool for education and training, especially in newer more autonomous modes for learning.

Acknowledgements

This project was supported by the UK Department of Employment Learning Technology Unit, Exeter University, ICL and Fujitsu, Microtek Ltd., Aldus Ltd., and Autodesk Ltd. We would also like to thank the staff and students involved at Exeter University, at St. Thomas High School, Exeter, and at Knowles Hill School, Newtown Abbot.

References

Cory, S. (1991). Technology in schools: who'll provide the leadership? *Computers in the Schools,* **8**, (1/2/3), 27-43.

Davis, N. E. (1993a). Implementing computer conferencing: from concept to participation. Presentation to *Computer Conferencing in Higher Education* conference, Oxford, July.

Davis N. E. (1993b). *Cost-benefit analysis of ISDN2 and DeskTop Conferencing.* Report to the Department of Employment Learning Technology Unit, Sheffield.

Davis, N. E. (1993c). Border crossings. *Times Educational Supplement,* 1/1/93, p.24.

Gale, S. (1992). Desktop video conferencing: technical advances and evaluation issues. *Computer Communications,* **15**, 5.

Littleton, K., Light, P., Messer, D., and Joiner, R. (1992). Collaboration and children's computer-based problem solving. *Poster presented at the NATO Advanced Study Institute on Psychological and Educational Foundations of Technology Based Learning Environments*, Crete, Greece.

Nardi, B. A., and Miller, J. R. (1990). An ethnographic study of distributed problem solving in spreadsheet development. *CSCW90 Proceedings*, Association of Computing Machinery.

Nordby, K. (1993). Human factors. In: *A Window on the Future: The Video Telephone Experience in Norway* (T. Kristiansen, ed.). Norwegian Telecom Research Department, pp 66-77.

Spears, R., Lea, M., and Lee, S. (1990). De-individuation and group polarisation in computer-mediated communication. *British Journal of Social Psychology*, **29**, 121-134.

von Tetzchner, S. and Holand, U. (1993). Supervision of rehabilitation and psychiatric work. In: *A Window on the Future: The Video Telephone Experience in Norway* (T. Kristiansen, ed.). Norwegian Telecom Research Department. pp 32-48.

Wright, B.W., Rogers, M., and Still, M. (1993). Desert Storm: a simulation. *Computer Education*, February.

Chapter 11

Co-Learn: an ISDN-based multimedia environment for collaborative learning

Tony Kaye

Introduction

Co-Learn is a research and development project in the current DELTA Programme, involving the design of an integrated groupware environment for collaborative learning at a distance, based on personal computers connected through ISDN links to a central host[1]. The Co-Learn environment contains several tools for real-time and asynchronous communication via text, graphics and speech, as well as tools for the preparation, sharing and annotation of multimedia courseware.

Co-Learn is one of the first of a new generation of prototypes which aim to integrate in one environment a range of tools and communication modes for collaboration between actors in distance education and open learning programmes. Two technological developments underlie this new generation of systems:

- easier access to relatively low-cost PC-based learner stations with multimedia software and hardware
- increasing availability of ISDN facilities, at least in resource centres, educational institutions and organisations (if not in users' homes), permitting simultaneous transmission of voice and data.

Use of ISDN facilities is essential for the full implementation of the Co-Learn environment for a number of reasons:

- for the transmission of high quality sound (7 kHz) via a digital audiographic bridge
- for the integrated use of the host server and audio bridge

---

[1]The members of the Co-Learn Consortium are Télésystèmes, Paris (lead partner); Laboratoire TRIGONE Université de Lille 1; Technische Hochschule, Darmstadt; Software de Base, Madrid; Universidad Politecnicá de Madrid; INESC, Universidade de Aveiro; IET, Open University.

- for input and retrieval of multimedia courseware items and mail messages with voice annotations from the database on the central host.

These functions, which contribute to the relative sophistication and complexity of Co-Learn compared to PC-based audiographic systems currently on the market, could not be easily supported, for example, through the simultaneous use of two PSTN lines. A potential disadvantage of reliance on ISDN is that learners and tutors need in general to go to centres specially equipped with ISDN facilities to use the system. In Co-learn, this is the case for any real-time sessions. However, users with PCs at home or in their office would be able to access the asynchronous forum in Co-Learn using a modem and a PSTN line. In this way, activities such as asynchronous group work, asynchronous online seminars or one-to-one electronic mail between learner and tutor, using a PSTN line, could be supported, providing a sense of continuity in between real-time, synchronous sessions over ISDN, as well as a chance to build on, and prepare for these sessions.

An important socio-cultural (rather than technical) influence on the development of the Co-Learn system has been a move away from transmissive and behaviourist models of education and training towards constructivist models, which place more emphasis on inter-personal collaboration, social networking, peer exchange and group activity in the learning process, and which recognise that learning cannot be considered in isolation from contextual and social factors. This movement is necessarily accompanied by changes in the traditional roles of both learners and tutors: learners are expected to take a more active and constructive role, contributing from their own experience and knowledge as appropriate; tutors are expected to be facilitators and resource people, as well as course planners and knowledge brokers. This change in emphasis is apparent in the school and university environment (Webb, 1982; Slavin, 1990; Duffy *et al*, 1993), within organisations (Senge, 1990) and in the distance education field, where it is sometimes referred to as 'third generation'[1] distance education (Nipper, 1989).

Co-Learn as a virtual resource centre

The Co-Learn environment can be considered as a virtual resource centre, with different 'rooms' which learners and tutors can visit (see Figure 11.1 on the next page).

---

[1]'First generation' distance education is represented by traditional correspondence tuition, and 'second generation' distance education is typified by multi-media mass education systems, such as the UK Open University.

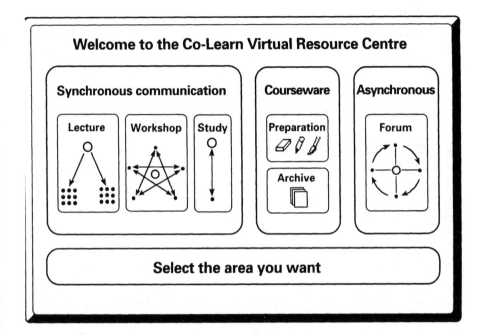

Figure 11.1   *The Co-Learn virtual resource centre*

The resource centre is in effect distributed between the user's workstation and the Co-Learn host server: the server verifies learner and tutor access, indicates which rooms are available for the course(s) in which they are involved and contains the courseware and asynchronous message databases. Tutors can set up specific subsets of rooms for their particular course and, within each type of room, different tools and resources are available.

Lecture room: real-time tele-teaching

This is based on the *Télé-Amphi* [1] audiographic application (Seguin *et al*, 1992), and is for remote lecturing from one to several sites, in conjunction with courseware (slides, diagrams etc) that has been downloaded to the learner workstations before the lecture, thus providing a remote 'electronic whiteboard' via local video projectors. The tutor can control the presentation on the remote screen and point at different parts of the

---

[1]Developed jointly by the ENSSAT and the CNET in Lannion, France.

screen. Learners can ask questions via an audio channel, turn-taking being controlled by the host computer.

Workshop: real-time multimedia conferencing

This is designed for a group of up to five participants, using the metaphor of a virtual meeting room. Each participant has a private space for documents and can move them into the public discussion space for synchronous voice discussion. One channel of the basic access ISDN is used for voice, the other for data. Where access to a broadband network is available, a video window would also be possible.

Private study: real-time tele-assistance

This permits direct connections between a central workstation and up to four 'slave' workstations. The tutor's central screen is able to represent the state of each of the learner workstations, and the tutor can help one learner at a time, or monitor up to four learners simultaneously.

Forum: asynchronous conferencing and electronic mail

This is a text-based computer conferencing tool specially adapted to the sorts of exchanges that occur in educational situations, with particular roles and types of message built into the design of the user interface. It should include possibilities for voice annotation of messages and for attachment of graphics and other files to messages.

Library: multimedia courseware repository

This is an object-oriented database on the host server, containing multimedia courseware objects (text and graphics with sound annotations) prepared by tutors with special Co-Learn editing tools. The library can be accessed from any of the four Co-Learn rooms.

Co-Learn as a research and development project

Co-Learn is a research study of computer support for collaborative learning, as well as a software development project, and over a dozen field trials in four different countries will be evaluated. A major focus of the evaluation will be on the reactions of users to a system which integrates in one environment a range of communication modes which, generally, are not found together. For example, a traditional computer conferencing system does not permit users to link messages to discussion objects (texts, diagrams etc) stored in the same database, and does not support real-time communication and collaboration within groups of users. A traditional audiographic system does not have a facility for asynchronous group communication between real-time sessions, let alone one which permits users to access and comment on the same courseware objects they have seen in a real-time session, at any time they wish. Tutors and learners in a number of the trials will already have experience of using existing audiographic and conferencing systems, and will therefore be well placed to judge the relative added value of a new system such as Co-Learn which tries to integrate several tools in one environment.

In addition, evaluation will address the social and cultural contexts in which the system is being used. An ergonomically successful integration of the various tools in the Co-Learn environment at the software and interface levels is no doubt a *necessary* condition for successful use of the system for collaborative learning, but it is not a *sufficient* condition. Many learners and tutors have little experience of the practice of collaborative learning, and even less understanding of the theoretical principles that underlie it[1], because much of their experience of education has been based on a transmissive model which places far more emphasis on individual achievement and mastery of pre-structured information, than it does on group synergy and active construction of knowledge. This is the case both in second generation distance education programmes, as well as in many traditional face-to-face, lecture-based courses. In interpreting the results of the trials, the evaluators will need to examine the extent to which tutors and learners in different pilots expect to use the Co-Learn system for promotion of interactivity and collaborative learning (as opposed to transmission of information and checking of understanding).

As Co-Learn integrates a number of different group communication technologies which have generally been used in isolation from each other in the past, findings from earlier studies of use of technologies to mediate human presence in distance education are briefly reviewed below, with a view to identifying the main research and evaluation issues to be addressed in the use of the Co-Learn environment.

Audioconferencing

The first experiences of the use of audioconferencing in education go back to 1939 when the Iowa School District (USA) started using telephones for provision of teaching to children ill at home or in hospital (Olgren and Parker, 1983). Since then, a large number of educational institutions in North America have used audioconferencing on a routine basis: a survey carried out in 1986 mentions over 50 universities with audioconferencing

---

[1]See, for example, Webb (1982), and Slavin (1990) for discussions of theory and practice of collaborative learning in the classroom environment; and Kaye (1992) for a summary of issues relevant to computer-mediated collaborative learning.

facilities (Bacsich, Kaye and Lefrere, 1986). The most important of these is the Educational Telephone Network at the University of Wisconsin-Extension, with over 200 reception sites, used by over 35,000 students every year. Most of the uses of telephone conferencing in North American universities are for remote delivery of classroom type instruction; however, some institutions, such as the Télé-Université in Québec, use the telephone for support and tutoring of students taking distance education courses (Abrioux, 1985).

In Europe, use of audioconferencing for education or training has been more limited than in North America, despite the existence in some countries (e.g. in France) of well-equipped audioconference suites and telephone conferencing services, widely used by many organisations for business meetings. In Britain, a national educational audioconferencing service (PACNET) was set up in 1984 at Plymouth Polytechnic; however, although the system was of an excellent technical quality, and used professional operators, it was unable to build up a big enough user base, and only lasted for two years (Winders, 1985). The British Open University makes some use of telephone conferencing for students in isolated areas of Scotland and Wales.

What can be learnt from these experiences? Firstly, it is clear that telephone conferencing can be used for group teaching, with results comparable to those found in face-to-face classes, for subjects as diverse as music teaching (Perenchief and Hugdahl, 1982), language teaching (Oksenholt, 1988; Kaye and Kerbrat, 1992) or in-service teacher training (Burge and Howard, 1990). In traditional distance education programmes, telephone meetings can be used with success for students unable to attend face-to-face tutorials (Robinson, 1991; Rekkedal, 1989).

Secondly, the basic principles which determine the success of learning in groups are the same whether the group is meeting face-to-face or by telephone, and a teacher who knows how to manage a face-to-face group successfully will soon become competent in running an audioconferencing group (George, 1990). However, the exclusive reliance on the audio channel, and the lack of visual cues, means that the teacher has to pay far more attention to the following aspects than in a face-to-face class:

- adequate preparation and training of participants in using the system, especially for the extra level of concentration required to take part in an audioconference
- very careful advance planning of the session, with a variety of different types of activity, and with different phases of the meeting clearly mapped out and supported by printed hand-outs and visuals which are sent to participants in advance of the sessions
- an opening phase of welcome and checking of each participants' presence and a closing phase for preparation of the next meeting and for informal chat

- constant attention to management of turn-taking, and the maintenance of a courteous and polite atmosphere in which no-one is left out, and no-one is allowed to monopolise the discussion
- in cases where one or more of the remote sites include a group of participants, the need to have one person act as local *animateur* and spokesperson for the group
- more frequent questioning by the teacher, to ensure that each participant is following the class and remaining involved; in general, audioconferencing is considered as being more appropriate for discussion-based and interactive 'classes' than for lectures and simple information delivery.

Finally, it should be stressed that it is generally technical issues and problems such as poor sound quality, malfunctioning of the audio bridge and uncomfortable listening conditions that deter users (e.g. handsets for individual participants are far less comfortable than headsets), rather than the fact of taking part in a group activity in the absence of eye contact and physical presence (Winders, 1985). This having been said, it is clear that there are some situations for which audioconferences, because of the lack of visual and body language cues, are probably inappropriate: these include initial sessions where participants are not familiar with each other, sessions involving conflictual, sensitive or highly complex issues, negotiation or bargaining situations and situations where a high degree of spontaneous interaction is needed. Many of these situations are ones in which physical presence is considered to be necessary, so these guidelines[1] most likely will apply to any type of teleconferencing environment.

Audiographic systems

Audiographic teleconferencing systems add to the audio channel various forms of low-bandwidth graphics, such as text, diagrams, telewriting or still pictures. The addition of the graphics element overcomes the obvious pictorial limitations of 'pure' audioconferencing, and some research has indicated that even in subjects with no obvious pictorial content, the use of graphics assists the teaching process, probably because the screen provides a focus of attention for the students.

The level, quality and flexibility in use of the visual support varies widely from one system to another, depending on the communication bandwidth available and the sophistication of tutor and learner stations, software and peripherals. At the lower end of the spectrum are systems such as the French TELECOURS product, which allows a tutor to send

---

[1]Derived from a *Program Manual on Teleconferencing* produced by the Knowledge network of British Columbia.

videotex pages to participants during a telephone conferencing session (students switching their phone line from voice to data during transmission); at the upper end are systems which use PCs or Macintosh computers, and ISDN lines to carry simultaneous voice and data signals, with users at different sites able to annotate graphics 'slides' using a keyboard or a graphics pen.

As with audioconferencing, there has been significant use of audiographics systems in education and training in North America for the last twenty-five years or so. However, early systems (such as the 'Electrowriter' and early 'electronic blackboard' systems) were fairly primitive, with low resolution black and white images, slow transmission speeds and no graphic feedback from reception sites (see, for example, Smith, 1992). In fact, at the University of Wisconsin, the systems that had been used for some time were abandoned in 1987 because of such problems and use of audiographics has only re-emerged as a success story since the introduction of a high-resolution PC-based system (the GTS system on the WisView network) using 9600 bit/s modems and twin telephone lines. At the UK Open University, use of the Cyclops shared-screen telewriting system (see Robinson, 1990; McConnell, 1984), which required two phone lines, was eventually abandoned for cost reasons and because of the reluctance of many students to go to local centres for the tutorial sessions.

Reports of the use of modern audiographics systems outside North America include experiences in Finland with a similar system to that now used at Wisconsin, in France with the *Télé-Amphi* system at university and post-graduate level (Guivarc'h and Seguin, 1992), 'tele-learning' classes in schools in Queensland, Australia (using Macintosh computers), facsimile machines and telephones (O'Grady, 1990), and the use of Electronic WhiteBoard (EWB) systems for undergraduate physics tutorials as a component of distance education courses at Universiti Sains, Malaysia (Idrus, 1992) and for a variety of vocational and university level courses in Scotland (Morris, 1991; Tuckey, 1992).

Before reviewing briefly the lessons that can be learnt from earlier research into audiographic conferencing, it is important to stress the context in which most uses are situated. Because of the equipment needed for audiographic communication, most experiences involve groups of learners clustered around a computer or a graphics pad, or collectively viewing a projected image of a computer screen, in a specially equipped centre (in contrast to many audioconferencing situations, which can easily involve individual participants using their own telephone). This group context – which may involve several groups in multi-site systems, including a group physically co-present with the teacher – introduces a fairly complex communicative situation, mixing telepresence with real presence[1].

In reviewing the results of earlier evaluations, firstly, it is clear that, other things being equal, the instructional effectiveness of audiographic classes or training sessions, as measured by achievement scores, can be equivalent to, if not better than, face-to-face classes (Chute *et al*, 1989). This effectiveness, however, seems to depend fairly critically on a number of conditions:

- *the need for the teacher to have highly organised classroom skills*, to be able to manage two or more groups simultaneously, as well as to control the use of the equipment effectively
- *the role of a co-ordinator/animateur/spokesperson at the remote site(s)*, whether this is someone formally appointed, such as a teaching assistant, or a 'leader' who emerges from the group naturally; this role combines organisational, technical and social responsibilities, and its importance varies depending on the size of the group and the pedagogical scenario; whatever the case, the role needs to be planned for and assumed
- *the quality of planning and preparation for each session*: here the same factors apply as with group-based audioconferencing sessions (making sure that all participants are present before the session starts, that they have a clear plan for the session, that there is a range of different types of presentation styles and activities within a session and that sessions are not over-long...)
- *the sound quality*, which is important for the teacher (the better the quality, the easier it is to pick up background noise cues from the remote sites), and for the students (poor sound quality and 'clipped' sound demands a level of concentration which distracts attention from the content of the trainer's or tutor's discourse); where there are large numbers of participants at remote sites, several portable microphones in the room are essential if audience participation is desired – half-duplex sound (as in 'press to speak' systems) is perceived as a major drawback by many users
- *the quality of the visuals used by the trainer or tutor*: the same guidelines for the preparation of good overhead projection transparencies apply, but even more so (size and density of lettering, well-produced graphics, careful sequencing etc); tutors thus need to be competent in the use of the graphics software for the system, or to have access to competent people to prepare their graphics for them

---

[1]Although desktop conferencing software for personal computers, aimed at individual users, is now beginning to appear in some advanced office environments, it is likely to be some time before it becomes used to any significant degree for training or educational applications.

- *good pointing and annotation tools* (underlining, selecting, highlighting, text annotation etc), preferably ones which can be used both by tutors to accompany their spoken commentary and by participants at remote sites for questioning, feedback or as an aid to answering questions posed by the tutor
- *good facilities for spontaneous graphics inputs during a session* (as opposed to pre-prepared graphics downloaded before the session), such as graphics tablets with different coloured pens for each site, light pens for writing directly on screen (thus eliminating the hand-eye co-ordination problem of graphics tablets and pens), keyboard input, graphics templates or the possibility of scanner input and/or parallel transmission of faxed documents.

The last two points, and the quality of the sound, are the most important factors in creating a feel of telepresence, to compensate for the lack of physical presence: if the tutor's voice can be heard clearly, even intimately (in the *Télé-Amphi* system, it sometimes seems that the tutor's voice is literally coming from a point 20 or 30 cm from one's ear), if the tutor can detect remote speakers' relative positions in the room, and if good use is made of pointing, underlining, drawing and telewriting during a session, then the remote person 'comes alive' and physical presence can somehow be felt as superfluous to the communication. Some reports even suggest that the absence of the distracting features of a tutor's physical appearance and mannerisms can help focus concentration on the message and the visuals, thus aiding assimilation and promoting learning effectiveness.

Cost studies of tele-training (mainly in a corporate setting) in North America have demonstrated that audio and audiographic training represent the most cost-effective methods of delivery, as compared to face-to-face training, programmed instruction, computer-based instruction and video broadcast systems (Chute *et al*, 1989). However, it is not possible to generalise to a wide range of educational situations from such studies: it is clear that the results would differ significantly according to the value placed on student/ trainee time and on travel substitution and opportunity costs.[1]

Computer conferencing systems

During the last ten years, there has been a rapid growth in the number of education and training organisations using asynchronous computer conferencing technology for various forms of online education. This

---

[1]See, for example, the cost calculations for various telematic training and education scenarios presented in the DELTA CCAM Deliverable on *Telematic Networks for Open and Distance Learning in the Tertiary Sector*, Volume 2. Heerlen: EADTU.

growth has been associated with increased access to personal computers and modems, as well as with the development of national and international academic networks (JANET, EARN, BITNET, the Internet etc.) which can provide low cost access to computer conferencing systems. Conferencing supports many-to-many communication (whereas electronic mail is designed essentially for one-to-one or one-to-many messaging) and conferencing software includes features specifically designed to help in the organisation, structuring and retrieval of messages. Messages can be linked to each other (e.g. as 'comments'), organised in different 'branches' or 'topics' of a conference and search commands can rapidly identify messages with particular keywords in their titles or in the body of the text. Support is also provided for tracking the activity of individual group members: for example, it is generally possible to see which messages in a conference a given member has read, or which messages have been written by a given member. Special commands are available to the person responsible for a conference (the 'moderator' or 'organiser') which can help in defining the membership of the conference, in keeping the discussion on track, and in scheduling the opening and closing of discussion topics.

Computer conferencing is being used in a variety of educational and training settings, for example:

- to provide totally online courses in a virtual classroom / virtual seminar context in which participants and tutor may only occasionally meet face-to-face
- as an adjunct to print or media-based distance education courses
- as an adjunct to face-to-face lecture-based programmes
- for development of special skills, such as writing or language learning
- as a basis for online games and simulations, for example in management training courses
- for organisational communication, informal learning, and information sharing, particularly in companies with widely distributed office locations.

A number of recent publications review and evaluate these applications, from a variety of different perspectives (see, for example: Mason and Kaye, 1989; Harasim, 1990; Waggoner, 1992; Kaye, 1992). One of the findings from such reviews is the extent to which computer conferencing is a communication medium in its own right, although sharing some of the features of spoken discourse (such as group interaction, and the possibility of rapid, spontaneous exchanges) and of written discourse (text-mediated, asynchronous, revisable). It has the potential to be an excellent medium for group work and collaborative learning, as all utterances are stored, retrievable and editable. Participants can contribute at their own pace, at times convenient to them, and the asynchronous nature of the medium allows time for reflection and careful composition of contributions. A variety of different forms of group activity can be supported through computer conferencing, including seminars, small group discussion, dyads, team presentations, debates, peer learning groups, and so on. For the potential of computer conferencing for collaborative learning to be fully realised, a number of factors seem to be important:

- The participants need time to become literate in this new medium: this includes not only obtaining an adequate mastery of the commands of the particular system and interface being used, but also mastery of the conventions and etiquette appropriate to a written, yet interactive, mode of communication, and a realisation that the quality and value of the course or programme is in large measure determined as much by their contributions as by those of the tutor or trainer.
- The online environment needs to be carefully designed before the start of a course or programme, but in such a way that it can be modified in response to the needs and contributions of the participants, and to the different types of group activity which evolve; the environment needs to provide appropriate 'places' (topics, conferences) for different discussion and seminar themes, and for different social and other purposes (e.g. a 'café' for informal socialising, a 'library' for long documents, and so on).
- The tutor needs to develop special skills in moderating conferencing discussions, to encourage participation and collaboration amongst participants (e.g. encouraging participants to build on each other's contributions, guiding the discussion, providing discussion summaries from time to time, and generally 'weaving together' the various discussion threads); the tutor also needs to recognise the limits of his/her control of discussion contributions: turn-taking can no longer be managed in the same way as in a synchronous face-to-face or audioconference situation.
- Group size can be critical, with numbers in the range 10 to 30 seeming to be optimal for many educational applications; there are often clear individual differences in contribution levels (with maybe one-third being very active, one third contributing occasionally, and one-third being passive 'lurkers', so a small group may not develop sufficient critical mass for discussion to take off); but individuals who are reticent in face-to-face discussions may be more active online, when they have more time for reflection, and do not need to compete for turn-taking.
- Although computer conferencing can offer cost and logistical savings over face-to-face classes, as can any other teleconferencing application, it does not provide any economies of scale, and there is evidence that teaching via computer conferencing can take more tutor time than an equivalent course taught face-to-face.

- Specific software and interface features can have an influence both on take-up of a conferencing system, and on the types of group activity and pedagogical scenarios that are supported (e.g. it seems to be easier to set up and run parallel discussions and small group activities in 'branching' systems).

Integration of computer conferencing with other delivery modes (face-to-face, print, broadcast etc.) in courses which are not entirely online and where conferencing is an adjunct method, needs to be given particular attention, with the specific functions of conferencing vis-à-vis other course components being made clear.

Computer-supported collaborative learning (CSCL) and CSCW

There is a developing body of practice and research in the field of computer-supported co-operative work (CSCW), which has led to an interest in applying relevant findings to structured computer support for (mainly synchronous) collaborative learning activities. Although asynchronous conferencing and messaging software is often classified as a category of CSCW system (see Rodden, 1991), the level of sophistication of computer support for inter-personal collaboration in most of these systems is in fact still fairly limited. More progress has been made in software support for synchronous groupware environments in mainstream CSCW research – for example, in design of multi-user editors, group decision support (GDSS) systems, co-ordination systems and desktop conferencing systems.

There are several key groupware /CSCW concepts (see, for example, Ellis *et al*, 1989; Bannon and Schmidt, 1991) which are relevant to the design of environments for promoting and supporting collaborative learning:

- *shared environment*: namely, a set of objects where both the objects and the actions performed on them are visible to the set of users
- *group window*: a collection of windows whose instances appear on different users' screens and are 'connected', such that an action performed in one instance (e.g. scrolling, or drawing a circle) makes the same action appear in the others
- *view*: a visual or multimedia representation of some part of a shared environment, with the possibility of different users (e.g. tutors, students, editors...) having different views, with varying levels of information or modes of presentation
- *role*: software support for specialist roles which might be taken on by different users, giving differential access to resources, tools, and

activities, and helping to structure and control the flow and exchange of communications between different roles

- *activity support tools*: group work involves a variety of different types of activity, and software has been developed to support generic activities such as brainstorming or structured discussions
- *co-ordination tools*: software for strengthening the sense of telepresence and for facilitating group task completion, for example by letting users view their own actions within the overall task context, informing them of the status of their own contributions, informing them of the status of other participants or generating automatic reminders of dates or times by which tasks should be accomplished
- *navigation tools*: tools such as 'browsers' for helping users to navigate inside complex virtual environments, providing information at all times on the three essential navigational questions (*Where am I? Where have I been? Where can I go?*).

Application of these concepts and tools to the development of environments specifically designed for collaborative learning is in its infancy, although a number of prototypes (in addition to Co-Learn: see Derycke and Viéville, 1993) have been developed and/or tested. Examples include a Macintosh-based system incorporating joint document creation, structured group interaction and role playing for a distance education course on *Renewable Energy Technology* (Alexander *et al*, 1993); a real-time computer-supported role-playing simulation for learning about production control and management systems (Vilers, 1992); and a PC-based system (*Simulta*) for real-time small group work and classroom demonstrations for computer science teaching (Cartereau, 1993). In comparison with the longer-established teleconferencing areas outlined above, there has not yet been sufficient experience of CSCL environments to be able to draw up generalisations about good and bad practice. Derycke (1992), however, has identified a number of issues which are important in developing the new generation of multimedia CSCL systems. These include:

- *navigation*: not only within a system during real-time sessions (e.g. moving from one 'room' to another), but in particular the selective browsing of earlier contributions to a discussion which are stored by the system, the handling of information selection and overload, and the constructive re-use of earlier messages in the development of a group discussion or a group task
- *group and activity structuring*: a virtual education environment needs to provide support for a variety of group sizes and activities, for people who will be occasional users of the system (as opposed to a heavily used office automation environment, for example) and there is a conflict to be resolved between providing a generic and easy to use

environment, or a set of more specialised environments for particular activities and pedagogical scenarios (debates, free discussions, seminars, simulations, case studies...)

- *integration*: both of different *media* (text, voice, pointing, telewriting) and of different *modes* of communication (real-time and asynchronous) and in particular, the design of a hypermedia interface which will help occasional users to move easily between different media and modes of communication.

Finally, it is important to point out that sound has been treated so far very much as a poor relation in CSCW applications; now that high quality sound is becoming a feature of personal computer systems, attention must be paid to handling spatialisation of sound, the functions of different types of auditory cues and the psychological role that auditory imagery might play in CSCL environments (Derycke and Barme, 1993).

Conclusion: conversation as the underlying paradigm

Successful collaborative learning is based on interaction and constructive co-operation between the various actors in the teaching/learning situation, based on a shared knowledge of earlier inputs to the collaborative process. Thus it could be said that the conversation represents the strongest underlying paradigm for any computer-mediated form of collaborative learning. A computer system can also be used to provide a basis for a collective group memory, because of the potential for storage and retrieval of certain elements of interaction (e.g. text messages and voice annotations in a computer conferencing system, successive drafts of documents and graphics etc.). This reinforces the strength of the conversation paradigm as does, in the case of Co-Learn, the existence of a voice channel in the three synchronous tools.

A computer-based communication system such as Co-Learn can also be used for storing and displaying courseware materials as well as the records of previous discussions. In applying the paradigm of the conversation to the collaborative learning situation, it is important to distinguish between these courseware items, that is, the objects of the conversation (texts, graphics or CBT material displayed in a shared workspace) and the utterances (spoken or written) which make up the elements of the discourse amongst the participants.

Conversations, however formal or informal the situation might be, and however chaotic and unstructured they might seem at first sight, seem to proceed as if they were regulated by a strongly held set of cultural and social norms. Summarising McGrath (1990), we can identify some of the main norms of the face-to-face conversational process within our own dominant culture as follows:

- only one person can speak at any given time
- there must be a 'default speaker' available at all times (maybe a group leader or tutor)
- during a discussion, the immediate prior speaker can exercise some control over who speaks next for a few moments after the end of their own contribution
- floor time is shared amongst all speakers, but not necessarily in an egalitarian manner
- lengths of pauses between contributions are critical, and must be neither too short (a too-short pause risks interruptions) nor too long (which would cause awkward silences)
- transitions between speakers are signalled by multiple (verbal, para-verbal and non-verbal) cues which are crucial for adjusting pause times and ensuring smooth transitions
- participants assume that there are no eavesdroppers or outsiders party to the discussion, and that there are no anonymous contributions
- participants assume that each input is in some way connected to an earlier input or that it foreshadows future inputs.

In discussions which are focussed on an educational purpose, the level of formalisation is generally much higher than in, for example, social conversation. The IRF (Initiation, Response, Feedback) pattern initially identified in school classrooms by Sinclair and Coulthard (1975), seems to represent an exchange structure which can be found in many situations from the primary school to the adult training seminar: the teacher initiates an exchange, a student responds and the teacher gives feedback. Such exchanges exist within a discourse structure which has particular characteristics. Tagg (1992), in building a 'heuristic model' of a typical face-to-face tutorial class, identifies three important characteristics of educational discourse:

- *temporal sequencing*: contributions are ordered by time
- *logical continuity*: contributions build on each other, with new elements referring back to earlier ones; hence participants in the discussion must have access to the totality of the preceding discussion, in order to be able to contribute to, and make sense of, other contributions
- *cognitive immediacy*: the creative vitality of a group discussion depends to large measure on the synergy and dynamism created by the continual formulation and re-formulation of participants' viewpoints as the discussion proceeds.

The design of the Co-Learn environment should reflect and support, at a generic level, the interactive and purposeful nature of educational discussion and exchange patterns. The precise nature of these patterns will vary from one educational situation to another, and from one Co-Learn tool to another, but attention needs to be given to identifying a set of basic collaborative learning activities which can be put together in different patterns corresponding to particular educational situations.

A major challenge in the design of the environment around a paradigm of conversation or discussion is that, in the face-to-face situation, the progress of a discussion depends on a host of meta-communicative acts, gestures and utterances (Watzlawick *et al*, 1972) and underlying control signals (Gensollen and Curien, 1985) emitted by participants, which are quite separate from the actual content of the discussion. In a mediated communication environment, which depends on audio and screen communication only, many of these meta-communicative and control signals will be lost or attenuated. The loss of some of these signals is likely to encourage participants – especially tutors – to opt for a more interactive, interrogatory form of communication (e.g. through frequent solicitations of responses from learners) than they would in a face-to-face class, as a device for monitoring the 'presence' of the learners in the session. Any tools that can be provided within the Co-Learn environment for monitoring group presence and activity, and for 'de-attenuating' meta-communicative and control functions, should be explored.

However, in designing the environment, and in developing pedagogical scenarios for its use, it is important to stress that the interactions supported through Co-Learn can never be seen as equivalent to a face-to-face tutorial, lecture or seminar. The environment promotes communication between actors, permits interruptions, allows for group work etc, but (even without the added complexity of synchronous *and* asynchronous communication) the patterns of these communications and interactions, and the social communication protocols and dynamics, will inevitably be different from those of a face-to-face setting. Users of Co-learn, and of other CSCL systems, will have to find new and different ways to interact, work and learn together.

Co-Learn is more like a group discussion or tutorial environment than a classroom or lecture-based environment, because of the variety of inter-personal communication modes that it supports, and that is why we are adopting the underlying conversational paradigm. However, this does not mean that participating in a tutorial via Co-Learn is the same as participating in a face-to-face tutorial. The communication characteristics are different: apart from the obvious features of overcoming location constraints (in synchronous mode) and time and location constraints (in asynchronous mode), many other factors define this difference, for example:

- Parallel processing is possible (with two or more participants making screen inputs at the same time; co-located participants might talk amongst themselves while simultaneously taking part in an electronic discussion with remote learners).

- There is less social control over the environment (it is easier to 'leave' an electronic room than it is to walk out of a face-to-face class; relatively anonymous contributions might be possible).
- View size is restricted (due to screen constraints), which may make it more difficult to maintain an overall perspective on the discussion, or to navigate through contributions – a well-known problem in courses run over computer conferencing systems.
- In audioconferencing mode, turn-taking and monitoring of individual presence has to be more rigorously structured than in a face-to-face discussion.

Educational discourse patterns and structures in Co-Learn will clearly differ from those in face-to-face classes, seminars or tutorials. An important aspect of the research and evaluation work to be carried out in the project will be the identification of these differences, and the analysis of the implications for the development of a third generation of interactive distance and open learning systems.

References

Abrioux, D. (1985). Les formules d'encadrement. In: *Le Savoir à Domicile: Pédagogie et Problématique de la Formation à Distance* (Henri, F. and Kaye A., eds). Québec: Presses de l'Université du Québec, pp 179-204.

Alexander, G., Lefrere, P. and Mathison, S. (1993). Towards collaborative learning at a distance. In: Pre-release papers for the NATO Advanced Research Workshop on *Collaborative Dialogue Technologies in Distance Learning,* April 24-27 1993, Segovia: Universidad Nacional de Educación a Distancia.

Bacsich, P., Kaye, A. and Lefrere, P. (1986). New information technologies for education and training – a brief survey. *Oxford Surveys in Information Technology*, 3, pp 271-318.

Bannon, L. and Schmidt, K. (1991). CSCW: four characters in search of a context. In: *Studies in Computer-Supported Cooperative Work* (Bowers, J. M. and Benford, S. D, eds). 3-16, Elsevier, Amsterdam.

Burge, E. and Howard, J. (1990). Audio-conferencing in graduate education: a case study. *American Journal of Distance Education*, 4, 2, pp 3-13.

Cartereau, M. (1993). A collaborative learning tool for computer science teaching. In: Proceedings of *TelePresence: First international Conference on Technologies and Theories for Human Cooperation, Collaboration, and Coordination.* Lille, March 22-24, 1993. University of Lille, Institut CUEEP.

Chute, A. G., Balthazar, L. B., and Poston, C. O. (1989). Learning from teletraining. In: *Readings in Distance Learning and Instruction* (M. G.Moore and G. C. Clark, eds.). Pergamon Press.

Derycke, A. C. (1992). Towards a hypermedium for collaborative learning? In: *Collaborative Learning through Computer Conferencing: The Najaden Papers* (Kaye, A. R., ed). NATO ASI Series, vol **90**: 211-224. Springer-Verlag, Heidelberg.

Derycke, A. C. and Barme, L. B. (1993). Some issues on sound in CSCW: an often neglected factor. Lille: Université de Lille 1, Laboratoire Trigone, CUEEP. (mimeo)

Derycke, A. C. and Viéville, C. (1993). Realtime multimedia conferencing system and collaborative learning. In: Pre-release papers for the NATO Advanced Research Workshop on *Collaborative Dialogue Technologies in Distance Learning*, April 24-27 1993, Segovia: Universidad Nacional de Educación a Distancia.

Duffy, T. M., Lowyck, J., and Jonassen, D. H. (eds) (1993). Designing environments for constructiuve learning. NATO ASI Series, vol **105**. Springer-Verlag, Heidelberg.

Ellis, C. A., Gibbs, S. J. and Rein, G. L. (1989). Groupware: the research and development issues. Austin, Texas: Microelectronics and Computer Technology Corp. (mimeo)

Gensollen, M. and Curien, N. (1985). De l'analyse du fonctionnement interactif à l'évaluation de marché des téléconférences. *Annales des Télécommunications*, **40**, 1-2.

George, J. W. (1990). Audioconferencing: just another small group activity. *Education and Training International*, **27**, 3, pp 244-248.

Guivarc'h, J. and Seguin, J. (1991). Rapport d'évaluation des cours et conférences par Télé-Amphi: juin 1990-juin 1991. ENSSAT, Lannion.

Harasim, L. (ed) (1990). *Online Education: Perspectives on a New Environment*. Praeger, New York.

Idrus, M. R. (1992). Enhancing teletutorials via collaborative learning: the Malaysian experience. *DEOSNEWS*, **2**, 14.

Kaye, A. R. (1992). Learning together apart. In: *Collaborative Learning through Computer Conferencing: The Najaden Papers* (A. R. Kaye, ed). NATO ASI Series, Vol **90**. Springer Verlag, Heidelberg, pp 1-24.

Kaye, A. R. and Kerbrat, C. (1992). Télécours d'anglais du tourisme: rapport d'évaluation. ATENA, Montpellier.

Mason, R. and Kaye A. R. (eds) (1989). *Mindweave: Communication, Computers and Distance Education*. Pergamon Press, Oxford.

McConnell, D. (1984). Cyclops shared screen teleconferencing. In: *The Role of Technology in Distance Education* (A. W. Bates, ed). Croom Helm, London, 139-155.

Nipper, S. (1989). Third generation distance learning and computer conferencing. In: *Mindweave: Communication, Computers and Distance Education* (R. Mason and A. R. Kaye, eds). Pergamon Press, Oxford.

O'Grady, G. (1990). *TeleLearning Project Report.* Department of Education, Queensland.

Oksenholt, S. (1988). A first-year college Norwegian class by telephone: an introductory comparison of methodologies. Eastern Montana College, USA. (mimeo)

Olgren, C. and Parker, L. (1983). *Teleconferencing Technology and Applications.* Artech House.

Perenchief, R. and Hugdahl, E. (1982). Teaching music by telephone. *Teaching at a Distance*, **22**, pp 33-34.

Rekkedal, T. (1989). The telephone as a medium of instruction and guidance in distance education. Research and Development Dept, NKI, Oslo.

Robinson, B. (1990). Telephone teaching and audioconferencing at the British Open University. In: *Media and Technology in European Distance Education* (A. W. Bates, ed). EADTU, Milton Keynes, 105-112.

Senge, P. M. (1990). *The Fifth Discipline: The Art and Practice of the Learning Organisation.* Doubleday/ Currency 1990, New York.

Smith, T. W. (1992). The evolution of audiographics teleconferencing for continuing education at the University of Wisconsin-Madison. *International Journal for Continuing Engineering Education*, **2**, 2/3/4, 155-160.

McGrath, J. E. (1990). Time matters in groups. In: *Intellectual Teamwork: Social and Technological Foundations of Cooperative Work* (J. Galegher et al, eds). Lawrence Erlbaum, Hillsdale, N.J.

Seguin, J., Guivarc'h, J., Morin, T. and Guillot, Y. (1992). Télé-Amphi: un service de communication de groupe. In: *L'Echo des Recherches*, no. 149, CNET.

Sinclair, J. M. and Coulthard, R. M. (1975). *Towards an Analysis of Discourse: the English used by Teachers and Pupils.* Oxford University Press, London.

Slavin, R. E. (1990). *Co-operative Learning: Theory, Research, and Practice.* Prentice Hall, Englewood Cliffs, N.J.

Tagg, A. (1992). Computer conferencing systems and their application: A suggested methodology. London: Birkbeck College, Dept. of Occupational Psychology. (mimeo)

Tuckey, C. J. (1992). Media for teaching and learning in distance education: computer conferencing and the electronic whiteboard. Edinburgh, Institute for Computer Based Learning, Heriot-Watt University. (mimeo)

Vilers, P. (1991). Etude d'un système de téléconférence temps-réel: application à un jeu de rôles en production. Lille: Université de Lille 1, Laboratoire Trigone, Thèse de Doctorat.

Waggoner, M. (ed) (1992). *Empowering Networks: Computer Conferencing in Education*. Educational Technology Publications, Englewood Cliffs, N.J.

Watzlawick, P., Helmick-Beavan, and Jackson, P. (1972). *Une Logique de la Communication*. Seuil, Paris.

Webb, N. M. (1982). Student interaction and learning in small groups. *Review of Educational Research,* **52**, 3, pp 421-445.

Winders, R. (1985). Teleconferencing: student interaction by telephone. *Programmed Learning and Educational Technology*, **22**, 4, 327-333.

Chapter 12

A multimedia environment for distributed learning

Peter Zorkoczy, Werner Steinbeck, Jochen Sandvoss, Thomas
Schütt, Clemens Dietel and Patrick Ardaens

Introduction and overview

ÉCOLE, école [ekol] n.f.
1(a) school; French word for school.
2 (a) Project of the European Community's DELTA program
to develop a Europe-wide system for distance learning; short
form for European Collaborative Open Learning
Environment

ECOLE is a multinational European Community supported project to design, implement and field test the prototype of a cost-effective distributed multimedia environment for learning purposes based on personal computers and digital communication networks. The project partners share a common view of three typical application scenarios underlying all technical work done in the project. This chapter reports on the concepts underlying the project, introduces the field tests defined, provides overview information on the technical design of the ECOLE workstation and presents some of the results obtained so far.

Goal of the project
The determining goal of the ECOLE project is

- to conceive, prototype and field-test
- a multimedia educational environment
- for distance learning
- based on available digital communications (LAN, ISDN).

The co-operating partners, internal project structure, technical system architecture, necessary applications, communication networks and field trials are all derived from this goal.

Project partners
Based on their specific areas of competence, three representatives from the European computer industry (BULL S.A., IBM Deutschland GmbH, Siemens AG) and seven European public network operators (France, Germany, Italy, Netherlands, Spain, Sweden, Switzerland) joined forces to conduct this challenging research project.

Project Structure
The ECOLE project consists of four main areas of work:

1 *field trials*:   Selection of representative learning scenarios; definition, co-ordination and evaluation of the field trials
2 *tool development*: The common user interface, the ECOLE-specific functions and the integration of existing applications to be used by field trials
3 *multimedia communication platform*:
                       Provision of isochronous transport services, point-to-point and multi-point connections based on existing network technologies
4 *technical and administrative co-ordination*:
                       Co-ordination of the technical work of the different work packages, internal communication and external relationship, European co-ordination, general administrative and management functions.

The technical interrelationship of the main areas of work and the contributing partners are illustrated in Figure 12.1. The areas of work will now be described in summary form. Greater detail will be found in later sections.

The field trials
The field trials were selected to define details of the three learning scenarios underlying the ECOLE project. They will test and validate the prototype being developed. Field trials will be carried out in different partner countries evaluating technical and financial aspects of the different learning scenarios.

The tools
Based on the field trial requirements, the ECOLE environment includes a variety of applications offering both traditional computer based training (CBT) programs and advanced services for collaborative learning.

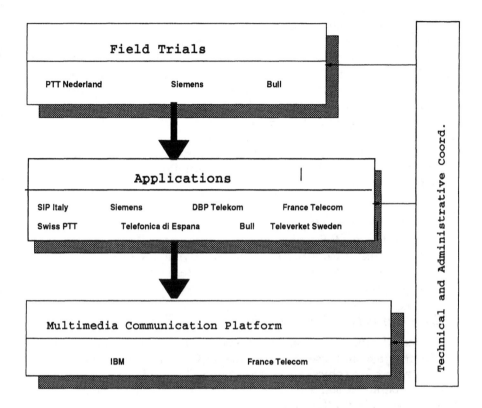

Figure 12.1 *The ECOLE project structure. Technical dependencies among the areas of work*

The Multimedia Communication Platform
The ECOLE Multimedia Communication Platform provides the basis for all distributed real-time multimedia applications. It offers isochronous transport services for multimedia data streams. The first Token Ring based version of the communication subsystem has been completed. Later versions will provide support for additional LANs, Basic Rate ISDN and later, Broadband ISDN.

Concepts for distributed learning

Learning and motivation for learning are influenced by a range of factors, which are not necessarily accessible to analysis, but are somewhere in the semiconscious or subconscious area of the human's cognition (Rogers, 1977). Social interaction is considered to be an essential component of the learning process and informal peer learning as important as formal teaching. Before using the ECOLE learning environment, it is necessary to know which factors influence the learning process. The ECOLE learning environment tries to simulate the well-known scenario of students and tutors interacting together in one room. It is unknown whether the character of the interactions will be sufficiently retained to induce the learner to look for the same benefits as in a face-to-face environment. Possibly, the character of the educational interaction changes into something new, more helpful to the learning process. It is therefore imperative to put more effort into the design of the facilities for communication and collaboration in the learning process than has been done for self-study.

Model of the learning process

The aim of the ECOLE project is to retain the benefits for the learning process that result from the interpersonal dynamics and motivational stimuli in collaborative learning while increasing the flexibility and efficiency of learning. Can this be achieved and if so, how?

Collaborative and multimedia learning: benefits and pitfalls

Multimedia learning materials are offered by different vendors on various platforms. Just as these systems, the ECOLE learning environment integrates text, graphics, audio and video. But, while traditional CBT systems operate in stand-alone mode (the user is on his/her own), the ECOLE workplace offers the valuable possibility of interaction with instructors and peers by multimedia communication.

The ECOLE learning environment combines traditional CBT with distance learning over a communication network and collaborative learning in groups. The main elements of the environment are:

Collaboration
In most distance learning (conventional or technology based) the student is in self-study mode. But work without a tutor can be ineffective. In contrast to existing (local and restricted) learning facilities like courses on CD-ROM or video-tape, the ECOLE learning environment focuses on educational settings where several learners come together for the mutual benefit of working together towards a learning goal. Communication is possible in any direction; learners can ask their tutors or talk to other learners.

Distance co-operation
The collaboration of the learners is independent of their geographical distribution; the participants communicate through appropriate

telecommunication facilities. The same is true for other co-operation scenarios which are not directly related to a learning process.

Deferred- or real-time communication

The possibility of communicating with co-learners and tutors in deferred time adds a new dimension to collaborative educational settings. If the learner is allowed to answer a question later, that is, not on the spot as in traditional classroom scenarios but after a period of time, a new type of learning process will evolve with benefits for some learning situations. The immediacy of the group situation will be diminished in favour of the opportunity to work on responses with more contemplation and depth.

Distributed data

Data such as learning material, videos, graphics, text files etc. need not be kept on the local computer, but are distributed on the network (this is transparent to the learner). All participants can have the same information, which will then be available and common for everybody. Due to the anticipated widespread use and the high commercial value of the courseware to be used, special care has to be taken to cover the aspects of user authentication. This must be sound enough to be a legal basis for later charging and data integrity – not only for the courseware, but also for various administrative data files.

Multimedia interface

Our usual way of communicating with somebody else is to talk to and see our communication partner(s). This natural way of interfacing will be provided by the ECOLE multimedia environment (using audio and video communication).

Social impact

Interactions in a classroom setting and the influence of these interactions on motivation, attention, comprehension and retention are also influenced to a considerable degree by group dynamics and the complexity of verbal and non-verbal communication (Jacques, 1991).

The interactions and signals passed through some communication systems will be filtered, changed, cut. Other forms of interpersonal interaction will appear. The question remains open in which way the change in the form people communicate with each other influences the quality of interpersonal communication.

To ease user acceptance and ensure the best possible learning results, ECOLE will be used not instead of, but in addition to conventional learning.

Economic impact

Skilled tutors are and will remain a scarce and expensive resource. So sharing a tutor as can be done with the ECOLE learning environment will lead to considerable savings in costs for education. There will be no more need for students to travel, which was for many a serious problem. There will be no need for a meeting place nor travel for both students and tutors. There will be no need for a conference room at a remote site, but something like a dedicated room for educational purposes with special equipment (ECOLE Multimedia Learning Environment). In some cases it might be more sensible to provide the necessary equipment at the desktop or workplace, so that the learner just needs to switch to the ECOLE software to take part in an educational event.

Field trials and learning scenarios

For the ECOLE project, the educational aspects of the field trials are the driving and determining part from which the needs of the applications and the transport service have been derived. To cover a broad range of possible scenarios, three field trials will show the educational effectiveness of the ECOLE learning environment. The following learning scenarios have been analysed for the ECOLE system:

Scenario 1   Virtual teams
Scenario 2   Distributed classroom
Scenario 3   Just in time training

The scenarios combine collaborative learning, learner-tutor relations and local CBT using applications on the PC, deferred- and real-time communications.

Scenario 1. Virtual teams
The virtual team scenario has been selected by the ECOLE partner PTT Nederland. It tests a project management course which is being adapted to the ECOLE environment and will make use of the ECOLE collaboration facilities. The participants are interconnected by both LAN and ISDN connections (Figure 12.2).

Today, the project management course is a 200 hour course with classroom training and distance learning based on text material. The ECOLE scenario envisages a mix of CBT, classroom training and a management game. ECOLE facilities will be incorporated into the existing course and its effectiveness will be validated by another group working with the conventional learning materials.

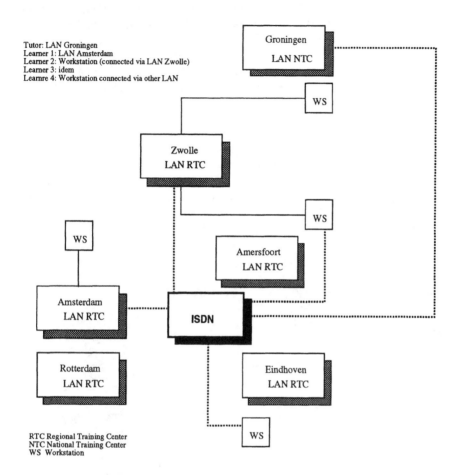

Figure 12.2 *Scenario 1. Training for project management.*
Field trial partner: PTT Nederland

Scenarios: PTT Nederland will link up six locations spread over the Netherlands, with up to ten workstations at each site.

Specification: Cost-effectiveness is a guideline for the Dutch field trial. Midrange PCs interconnected by LANs (Ethernet) and ISDN for the modular training program will support the ECOLE learning environment. The program elements of this field trial contain modules on different topics, consultancy, and applying the acquired skills in the context of a management game. The distribution of the learning material will be done electronically; self-study will be supported by local tutors, collaborative work and conferencing among the learners will be done using the ECOLE communication facilities.

Application programs: The requirements for this field trial include a range of communication facilities (file transfer, electronic mail, real-time and deferred-time communication facilities, audio and videoconferencing). Schedules have to be provided together with a question and answer facility supported either directly with a tutor or indirectly via a database. For pedagogical reasons, learners will be organised into groups.

Scenario 2. Distributed classroom
The distributed classroom scenario is the result of a collaboration of the two French partners in the ECOLE project, Bull and France Télécom. The field trial combines classroom activities and self study. Learners will be connected via LANs and/or ISDN. The field trial scenario follows an existing course currently based on paper and local CBT courseware about office automation. The ECOLE learning environment is intended to increase the interactivity of conventional distance learning.

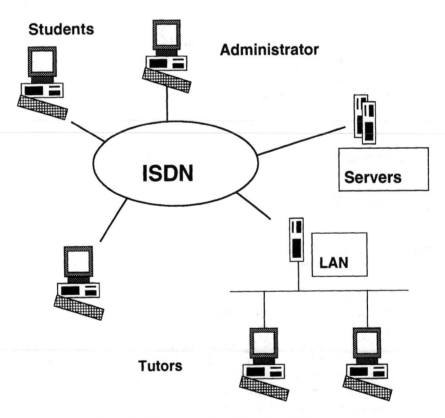

Figure 12.3   *Scenario 2. Learning for office automation.*
Field trial partner: Bull/France Télécom

The French field trial will use the following services offered by the ECOLE learning environment:

- Distribution of learning materials
- Communication among the students (collaboration)
- Communication with the tutors
- Access to a remote database machine storing learning material and administrative information
- Tutoring and monitoring tools to support and evaluate the collaborative learning process.

Scenario 3. Just in time training
The third scenario and field trial have been defined by Siemens. They intend to test ECOLE's effectiveness for just-in-time training of maintenance staff.

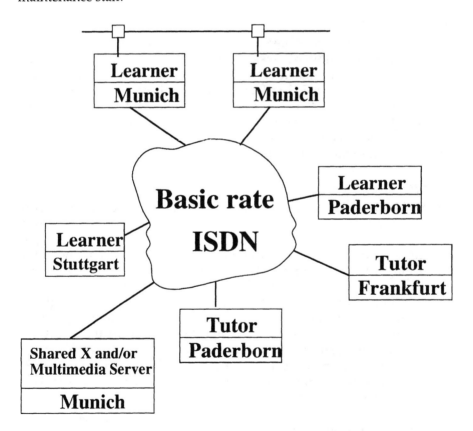

Figure 12.4   *Scenario 3. Product training for technical maintenance staff. Field trial partner: Siemens*

When new products emerge on the market, problems and questions concerning these products arise. Maintenance staff members need to be trained in time to deal with these. The conventional solution of providing CBT programs does not sufficiently solve the problem. The lack of skilled trainers and the effort of producing new CBT courseware that should foresee all the questions that could arise during its study made the responsible persons think about other solutions. Siemens wants to solve the problem by a combination of CBT, remote tutoring and real-time access to remote learning material. The participants are connected via ISDN (or LAN) and located all over Germany (Figure 12.4).

The learners will be able to use real-time video communication facilities and deferred-time communications like electronic mail. Database access and the possibility of sharing documents and doing joint editing will be provided by the ECOLE learning environment.

ECOLE services

The ECOLE services are offered to the user via the ECOLE user interface. According to the different tasks of the users in the different ECOLE scenarios, we have designed user interfaces with various functionality. The design of the ECOLE User Interface has been specified in an ECOLE Deliverable (ECOLE, 1992). It follows the principles laid down in the IBM Common User Access Advanced Interface Design Guide (IBM, 1989).

A more detailed description of the user interface can be found in the Appendix at the end of this chapter.

Learner's view

The ECOLE User Interface will combine existing (or basic) facilities with ECOLE-specific collaborative functions. The basic functions are those provided by traditional learning systems (editor, courseware, catalogue functions, local hypertext and multimedia applications, etc.). The point that distinguishes ECOLE form other systems is the distributed classroom situation, which requires special tools for group situation control.

The ECOLE User Interface offers a common look and feel throughout the applications to ensure easy access to the system and a positive reinforcement in its use. A learner using the system for the first time should be able to use the system successfully within a quarter of an hour.

The ECOLE User Interface has to be also flexible enough to integrate a variety of services including video communication and advanced services. The result is the ECOLE Shell and the ECOLE Screen.

Beyond the learner interface

The ECOLE project focuses on learners and their needs, but it is not forgotten that tutors, authors and administrators participate in the learning process and also have to interface to the ECOLE system. There are several modes envisaged to support their work:

• Setting up the ECOLE system	(administrator's view)
• Conducting a tutorial session	(teacher's view)
• Lecture delivery	(author's view)
• Conducting a real-time seminar	
• Conducting a deferred-time seminar	
• Designing and editing courseware	(author's view)
• Designing/editing educational events	(administrator's view)
• Evaluation of the educational event	(tutor, author, administrator)
• ECOLE tutorial	(everybody).

To support these modes, more facilities may be included, such as:

- Editor for a database containing the educational events
- Management facility for organising the educational events.

Applications and tools

As has been said above, ECOLE provides the technology necessary to demonstrate the educational effectiveness of telecommunications and computer based collaborative learning by field testing. In technical terms, the ECOLE environment consists of the applications, the computer system software and hardware, and the telecommunication facilities plus their software.

The section entitled Learner's View, was dedicated to the user's view of the applications. This section deals with the system view of the applications, the computer technology and the telecommunication facilities.

System view of the applications

The applications can be divided into communication-oriented applications and local applications. The latter only make use of local resources (local disk or CD-ROM, local printer, other multimedia input/output devices etc.). The other applications communicate with other computers by the telecommunication facilities. These include deferred-time and real-time applications. The deferred-time applications include:

- *Electronic mail*: The well-known facility of message sending (letters, files, graphics,...) together with a multimedia attachment (e.g. voice announcement, still image).
- *Electronic conferencing*: Users can share one or several common data areas used as a kind of pin-board, where everybody can leave a message.

- *File transfer*: Including texts, graphics, digitised still images, digitised audio and video can also be sent as plain file.
- *Database access*: The access to a remote database machine can be considered as deferred-time communication, as long as no time-critical data is to be transferred. The scope of the database is outlined below.
- *Distribution and updating of learning material*: Deferred-time communication is also needed for distributing files, courseware etc. needed by the participants.

The following applications need real-time communication:

- *Electronic conferencing*: The real-time version of electronic conferencing, as planned for the ECOLE prototype, offers audio and video communication with all participants.
- *Joint viewing/editing of a document with voice discussion*: This application needs simultaneous transfer of audio (isochronous streams) and data.

The educational database has three sections:

1 *Educational materials*: The database contains (i) topics which can be searched for by different keys (keywords, type, topic,...) and (ii) questions and answers.
2 *People*: The people profile in the ECOLE database can contain data such as name, phone number, electronic mail address, picture, educational background, access rights etc.
3 *Schedules*: The schedules in the database can be accessed by dates, material or instructor selection, type of educational event or other criteria.

The Multimedia Communication Platform

As already described above, the usage of different media in CBT can significantly improve the learning success. The core of the ECOLE System is therefore a multimedia communication platform which allows a distributed multimedia applications to be built on top of it. The ECOLE Multimedia Communication Platform enables not only standard data transfer but also real-time audio and video communication.

A variety of machines are being used for dedicated purposes:

- Individual workstations
  - Different groups of users (teachers, authors, learners, administrators,...)
  - Variety of application software
  - Multimedia equipment
  - Isochronous communication facilities

- Common communication subsystem, common operating system, common look and feel
- Database machines (Remote access to multimedia material)
- File servers
- Shared applications
- Gateways (architectural provision) for LAN and ISDN connectivity.

The heterogeneity of the different networks used in ECOLE is transparent to the user. For all the underlying networks, a common high-speed communication subsystem provides identical transport services.

The Multimedia Communication Platform is tightly coupled to the underlying operating system, the hardware components and the underlying network technology. All these components are described below.

Hardware and operating system platform

- End user workstations: 80386 and upwards
  - with full audio/video capability: OS/2 2.0 and MS Windows (WIN-OS/2)
  - without full audio/video support: native MS Windows running on DOS
- Servers: 80486 or RISC (UNIX).

Depending on the purpose of the machine, different resources will have to be made available: RAM, disk space, audiovisual attachments, network adaptors, loudspeakers, microphone, printer etc.

Networking technology

The telecommunication networks used in ECOLE include:

- Local Area Networks (Token Ring 1992, Ethernet 1993)
- Basic rate ISDN (1993)
- Broadband ISDN (1994).

The Multimedia Transport Service

The multimedia transport system of ECOLE (ECOLE, 1992b) offers its users standard data transfer services as well as an isochronous transport service for multimedia data. These services are necessary to support real-time communication as used in virtual classroom scenarios with multimedia collaboration or video distribution applications.

The service interface allows the application to adjust the quality of service (QOS) offered by the transport system to its needs. The application can choose whether the connection will have guaranteed properties or whether it is managed in best-effort style. It is up to the application to decide whether throughput or delay should be optimised, to determine the TSDU (Transport Service Data Unit) rate, the time gap between TSDUs, their length and to define a work-ahead (the number of TSDUs that is allowed to be sent above the negotiated rate). The application specifies the maximum acceptable error probability in terms of late data, lost TSDUs or corrupted data.

Besides these points the possibility of opening multi-destination connections (multicasting) is an important feature in the ECOLE learning environment where data is very often transmitted from one teacher or server to a group of learners.

The Transport Service Interface

The ECOLE transport service interface offers an open application programming interface to its users. In order to provide high performance data transfer services even on future high speed networks the interface is following the upcall concept where the control flow direction is directly mapped onto the data flow direction. Besides the downcalls that can be expected from the transport service interface (e.g. initialisation, opening a service access point (SAP), sending of data, connection establishment, wait for incoming connections, accept connection, refuse and release connection) incoming events (e.g. triggered by incoming data at the receiver side) are handled by upcalls.

Structure of the Transport Subsystem

The transport system consists of three main parts plus a run-time environment. OSI layer 4 is filled by the transport protocol which offers a flexible error handling. It relies on the ST-II (stream protocol II), which is a member of the TCP/IP family. ST-II is a network layer protocol and offers the possibility of multi-party connections, stream handling and resource negotiation. Below this protocol lies the interface to the network adaptors. It consists of two parts: the upper layer which is device-independent and manages the communication between ST-II entities and the drivers, and the device-dependent software (drivers depending on the sub-network). All the ECOLE multimedia transport system software will run on top of different kinds of networks (LANs, ISDN, B-ISDN) except this small part dealing with the network adaptors, which can not be realised independently of the underlying hardware.

Besides the communication protocols, several support instances have been implemented, in particular a buffer management system (BMS) for efficient data transport within the local machine (from network adaptor to transport system to application), a resource management system (RMS) dealing with the quality-of-service parameters negotiated during connection establishment and an operating system shield (OSS) for high portability.

Dynamic behaviour of the Transport Subsystem

Inside the transport subsystem the handling of control messages and data messages are completely separated. The handling of control messages is done by one common carrier of activity[1] for all connections. The handling of data messages is mapped onto one separate carrier for each connection. If the application establishes a connection, it calls the appropriate API functions.

Inside the transport subsystem, a carrier of activity is started. With every connection established, another carrier of activity will be started which runs as long as the connection is alive. At the receiver's side, a carrier of activity is waiting at the network interface for incoming data. If a packet arrives, a dispatcher ensures that it will be handled by the carrier of the corresponding connection.

Heterogeneity

The ÉCOLE Transport system runs on different operating system platforms. Currently versions are developed for OS/2, WINDOWS (WIN-OS/2) and AIX. The protocol implementation is designed to support regular personal workstations, client-servers scenarios and gateways.

Summary

ECOLE addresses the problem of integrating real-time and deferred-time interaction and group communication into a Europe-wide multimedia distance learning service. Although the project is essentially a technical one, its success will be measured by its contribution to the effectiveness of training.

The integrated prototype will include:

- an ISDN-based telecommunication network interconnecting both individual workstations and groups of workstations connected to local area networks
- telecommunication software, including an Application Programming Interface, allowing tools and applications to access digital telecommunication services
- software tools and applications to support real-time and deferred-time interaction and group working (learning, tutoring, authoring)
- a unified desktop environment giving convenient access to the tools and applications.

The prototypes will be tested and evaluated in company training environments, in three contexts:

---

[1]'process' or 'thread', depending on the operating system platform

- management training
- distributed classroom (dealing with office automation)
- just-in-time training of maintenance staff.

It is intended that the prototype environment developed in the project will subsequently form the basis of products and services in the field of technology supported training.

Workplan
The ECOLE activities are conducted in five phases. The detailed schedule is given in the following list.

PHASE I: System architecture, Communication, application, tools and interface specification (1992)

PHASE II: Multimedia communication system for LAN, early platform for application and tools development (1993)

PHASE III: Basic Rate ISDN integration, distributed applications (1993)

PHASE IV: Start of field trials (1994)

PHASE V: Broadband WAN (B-ISDN) integration (1994).

Appendix. Requirements for the ECOLE User Interface

The ECOLE User Interface has to be:

- user-oriented
- consistent
- easy to use
- based on a graphical user interface.

The design of the User Interface follows the principles laid down in the IBM Common User Access Advanced Interface Design Guide.(IBM, 1989).

An ECOLE learning session
Learners identify themselves and sign onto the system. With the learner's identification, privileges and categories will be set. Entering the system, the user is prompted by the ECOLE SHELL. This is the starting environment for all applications of the ECOLE Learning Environment. The learner can now access a mode (learning environment) or request help. The learner can access one of four educational settings (environments, modes) from where the different ECOLE facilities can be called. The modes are:

- Private study
- Supervised group learning
- Message oriented seminar
- Educational planning.

The main ECOLE facilities offered to the user are:

- Real-time communications: joint editing, audiovisual communications
- Deferred-time communications: messages, file transfer
- Database access: learning material, schedules, question and answer files
- Query one's own status: courses taken, credits, billing information etc.
- Recording of learning events, taking notes for personal use
- Useful small applications (tools): Diary, Clock, Calendar, Calculator,...
- Help function.

The ECOLE screen

The ECOLE User Interface is based on the WIMP paradigm, as are all the modern user interfaces. The user accesses the system by using windows or icons or menus moving through the system with a (mouse) pointer.

The main elements of the ECOLE screen are:

- Menu line
- Quick function bar
- Status bar.

Menu line

The menu line offers the following options:

- **File**   Local file management
- **Edit**   Basic editing functions
- **Mode-Specific**   Functions depending on the operational mode that are not time-critical.
- **Communications** Real- and deferred-time communication, group communication
- **Database**   Access to the educational database
- **Toolbox**   Providing small applications
- **System**   Mode-specific configuration and screen customisation
- **Help**   The Help function (hypertext facility).

Quick function bar

The function bar offers the ECOLE user fingertip access to the most important functions. The services offered here are dependent on the operational mode, but include the following:

- Navigation and browsing functions
- Services for group situation control
- Record function
- Print function.

The status bar
The status bar serves several purposes:

- Status line
- Command line.

References

Grebe, H. (1992). The ECOLE user interface: user environment and facilities. *Interface Concept Report DELTA Project D2004*: ECOLE, WP B2.1, Darmstadt.

Mulder, J. (ed) (1992). Service requirement document ECOLE field trials. *Final Version DELTA Project D2004*: ECOLE, WP D1.

Sandvoss, J. (1992). ECOLE: transport service interface. *DELTA Project D2004*: ECOLE, WP B1.2, Heidelberg.

Walker, R. and Schütt, T. (1992). ECOLE technical audit report, Brussels, October 14, 1992; Rapperswil, Heidelberg.

IBM. (1989). International Business Machines Corp. systems application architecture: common user access – advanced interface design guide IBM Corp., Doc. No. SY0328-300-R00-1089.

Jacques, D. (1991). *Learning in Groups*. Kogan Page, London.

Rogers, J. (1977). *Adults Learning*. The Open University Press, Milton Keynes.

Zorkoczy, P. (1993). Educational scenarios for telecommunication applications Proc. Advanced Research Workshop on *Collaborative Dialogue Techniques in Distance Learning*, Segovia, 1993.

Chapter 13

The integration of ISDN into a multi-functional visual communications network

Richard Beckwith

Introduction

The advent of ISDN as a new telecommunications service has brought many opportunities for concomitant applications. The most obvious of these applications, and perhaps the simplest idea for PTTs on which to base marketing, is the combination and exchange of data and voice in a two way business telephone call. In this instance ISDN can be viewed as a stand-alone service offering greater versatility (albeit at a higher tariff) than the Public Switched Telephone Network (PSTN). Other forms of stand-alone use of ISDN include digital link fall-back in the case of main link failure in computer networks and bandwidth-on-demand applications in which digital link capacity can be varied according to network loading.

The application of ISDN described here is essentially complementary to, and extending the functionality of, a broadband network (LIVE-NET) designed for interactive visual communications. The availability of ISDN enables a new form of outreach from this 'island' network both nationally and internationally which could not be realised otherwise because of technology and cost constraints.

Overview

In this chapter we will first consider the nature of visual communications, interaction, and technology options for implementation. We will then review the aims, architecture and applications of LIVE-NET (the London Interactive Video for Education Network) and the outreach of this network through satellite and ISDN. A particular application described is the Image Server: a shared resource with access from ISDN for image loading, retrieval and interactive browsing. We next consider models for multi-site interaction including the integration of ISDN links into LIVE-NET and satellite sessions. As a further development, a hub model with LIVE-NET functionality for the interworking of ISDN-connected videoconferencing terminals is proposed. In conclusion, scenarios for the future migration of the LIVE-NET broadband network to an all ISDN on-demand network are suggested.

Visual communications

Concepts

The term 'visual communications' is used here to convey a wider range of applications than are normally associated with the term 'videoconferencing'. Videoconferencing using compressed digital techniques has been marketed primarily as a replacement for face-to-face business meetings where cost benefit analysis can demonstrate the effectiveness of using technology to achieve meetings at a distance. This technology previously required expensive fixed digital circuits, high cost codecs and purpose designed rooms. More recently the marketing of videoconferencing has been extended successfully to include educational applications, as network and equipment costs have reduced.

In the wider context of visual communications, we are seeking enabling technologies that can transmit full motion television images without impairment and with a high level of interaction between multiple locations simultaneously. The applications that can be supported by such a network should not be limited by the characteristics of a more restrictive technology which might result in impairments such as a motion loss, low picture resolution, tonal compression and loss of colour fidelity. The educational value of such a network is that any visual material that could be used as part of a television broadcast or closed circuit transmission (for example from an operating theatre) can be carried by the network.

Realisations

There are two routes by which the enabling technology for a terrestrial visual communications network could be realised. The first option is in the analogue domain using broadband cable television technology. Both AM (Amplitude Modulation) and FM (Frequency Modulation) systems have been developed for multi-channel operation over fibre-optic cables. The AM systems can carry a far greater number of channels due to the minimal requirements for channel spacing. FM systems carry fewer channels with wider channel spacing but with higher quality at greater range.

The second option is in the digital domain using wide-band circuits and codecs (COder/DECoders) to convert analogue television signals to digital data. The data rate equivalent of a broadcast analogue television signal is 140 Mbit/s (megabits per second). Using a high bit rate codec, a near broadcast quality image can be achieved at 34 Mbit/s. codecs designed for videoconferencing use employ advanced picture processing techniques to generate images similar in quality to a domestic video cassette recorder, yet at much lower data rates in the region of 384 kbit/s (kilobits per second) to 2 Mbit/s. For low bandwidth applications over ISDN, videoconferencing terminals can produce usable images at 128 kbit/s.

Interaction

Another consideration for the enabling network technology is interaction which in simple terms is the ability for two locations to be in simultaneous visual and audio communication. If we wish to have more than two locations involved in simultaneous interactive communication how might this be achieved?

Within the analogue domain, cable television technology provides a 'channel rich' environment at a relatively low cost per channel. By providing each location with access to multiple channels, images of several sites can be conveyed simultaneously thus creating an ambience of 'full telepresence' – the ability for each location to see, hear and interact with every other location.

The cost of channel provision within the digital domain is considerably higher and determined mainly by the cost of the codecs employed. As a result, channel provision is normally limited to one duplex (transmit and receive) channel to each location in the network. Interactive communication between several locations is achieved through the combination of an open sound system and picture switching which is either voice activated or under keyboard control from a chair location. In videoconferencing networks the sound and picture switching functions are handled within a single device usually termed a Multipoint Control Unit (MCU).

LIVE-NET

Origins and aims

In 1985 the University of London, working within a framework for new communications technology, entered into a joint initiative with British Telecom (now BT) to develop an interactive visual communications network (Powter and Beckwith, 1988) to link five of the major Colleges of the University and two service centres. The objectives of the network were set out in two phases.

Phase one sought to establish the network and provide interactive real-time video links to multiple sites for teaching, research, demonstration and student activities. The network aimed to maximise opportunities for collaborative working and to enhance communications between teachers and students at different sites, through the provision of new audio-visual facilities for live lecturing and tutorials. The network would feature automatic modes of operation for channel configuration, and control of audio-visual resources would be by the users directly.

Phase two objectives were to serve as a focus for research and development in the whole area of related communications technology from electronic engineering and computer software control to studies in human factors and ergonomics. The image server described later is an example of the second phase development of LIVE-NET.

Architecture

The enabling technology for LIVE-NET had been developed by BT initially for the Westminster Cable Company franchise and was the first use of a fibre-optic based analogue broadband FM cable television network in the UK. In order to meet the requirements for the granting of a franchise licence the network intrinsically had to support interactive video return channels. This was achieved through the use of a switched-star architecture with head-end distribution to a tier of main hub switches and further distribution to local wide-band switch points serving up to 48 customers. Return video channels went from each switch point to the hub switches and then back to the head end. Customers of Westminster network had direct control of their local switch point for programme channel selection.

For LIVE-NET, the transmission and switching components were re-deployed to create a symmetrical network with equal transmit and receive channel capacity to each site. The FM broadband transmission system carries four near-broadcast-quality television channels over multimode fibre for links up to 3 km, and monomode fibre for longer links up to 25 km before a repeater becomes necessary. The video channels to and from each site terminate on a central hub switch derived from the Westminster video library switch architecture. This is a 64 x 64 non-blocking cross-point switch with multiple 1:1 and 1:n input to output assignment capability. Control of the switch is via a serial communications interface using a message passing protocol. In addition to the four video channels (with sound on a sub-carrier) each fibre also carries a fifth analogue channel with digital signalling at 2 Mbit/s for network control and data transfer.

Implementation

The LIVE-NET fibre-optic cable network, transmission equipment, switching and site terminations were fully commissioned by BT in December 1986. The network linking the five major Colleges of the University of London (Imperial College of Science, Technology and Medicine, Kings College, Queen Mary Westfield College, Royal Holloway and Bedford New College and University College, and two service centres: University of London Computer Centre and the Audio Visual Centre) extended from the East End of London to Egham in Surrey with a total of 62 km of fibre laid in BT ducts. Whereas only multimode fibre had been used previously in the Westminster network, LIVE-NET was the first installation of the broadband monomode equipment developed by the BT Research Laboratories at Martlesham.

The role of the LIVE-NET group in the Central University, in collaboration with the connected sites, was to develop the software switching control of the network. It also aimed to liaise in the design of the end user equipment for lecture theatres and seminar rooms to provide live access to the network. The switching system runs in conjunction with a database to automatically control the setting up of channel configurations for sessions booked throughout an academic year. Subsequently the switching system was extended to control, in addition, smaller (32 x 32) video cross-point switches installed at each of the major sites for the interconnection of multiple audio-visual facilities via local video networks to the main network. For example, University College London (UCL) has links to two hospital sites (Middlesex and Whittington) which can be interconnected to the main UCL site and/or linked to the main LIVE-NET network for, *inter alia*, the teaching of surgery. A diagram showing the present network including the link to the University of Westminster is shown in Figure 13.1.

Functionality

The combination of multiple television channels and computer controlled switching enables LIVE-NET to achieve a wide diversity of functional configurations. At the simplest level, a video input channel from one site can be programmed to be sent to one or several others in a broadcast mode. This might be used to relay a video tape or to transmit a live event to a satellite uplink. The essential aim of the network, however, is to achieve interactive visual communication between sites.

Since each site has four transmit and receive video channels to the hub switch, a session configuration mode termed 'full interconnect' is used frequently. In this mode each site transmits one channel to the network and receives back the images from every other site in the session on multiple receive channels. With four receive channels available at each site, a total of five sites can be interconnected in this way with simultaneous vision and sound. At each site each receive channel is displayed on a separate television monitor. Since sound is carried with the video signal, this results in a spatial separation of the sound sources aiding the identification of the speaking site.

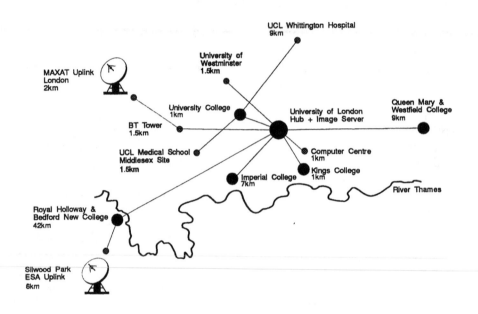

Figure 13.1　*LIVE-NET Fibre-optic Network*

The full interconnect mode is ideal for small group meetings and seminars where a high level of interaction is needed. It is also well suited to sessions involving new users of the network since no control of picture switching is necessary. (More complex configurations involving in-session channel switching are described in a later section). For sessions involving large numbers of students in lecture theatres, the images of the lecturer and visual material are more important than students seeing their fellow students simultaneously at other sites. In this configuration all of the student sites are seen simultaneously by the lecturer on separate small monitors or multiplexed on to a single monitor. The student sites see the lecturer and visual material usually on a large screen projection television system.

Each LIVE-NET room is normally equipped for both lecturing and student audiences. The speaking position has a video camera for the lecturer/presenter, an overhead camera for visual aids, a slide-to-video convertor and access to other sources of video input, for example, a video cassette recorder, computer graphics and laser disc player. The selection and control of all the video sources is controlled by the lecturer/presenter directly without technical assistance.

In meeting situations, one of the attendees takes responsibility for video switching and camera control at each site. The speaker cameras are equipped with remote pan/tilt and zoom controls for speaker selection within the group at each site. Because several sites can be interconnected in a session, control of cameras other than at the local site is not thought desirable and has not been implemented. Local camera position selection using shot box controllers has been experimented with but the cost of these controllers, compared with the decreasing cost of CCD video equipment, makes multiple cameras a more economic proposition. In one LIVE-NET room, for example, three speaker cameras, each covering two to three attendees (Figure 13.3), can be zone switched instantly.

Applications

The main application of LIVE-NET is the teaching of undergraduate and postgraduate students who are registered for courses that are being taught on an inter-collegiate basis within the University of London. Many of the undergraduate courses are in the humanities (for example history and classics) where the ability to take a wide variety of options makes the courses particularly attractive to students. A significant factor is the interest shown by lecturers in providing far more visual material than would have been the case with conventionally delivered lectures. This is mainly because of the ease of showing any material that may be to hand, including photographs, pictures from books, post cards and artefacts, by means of the overhead video camera.

In the sciences post-graduate courses have been given each year in quantum electrodynamics where the ability to develop the construction of complex equations and zoom into detailed expressions again using the overhead camera, has proved particularly effective. Other subject areas have included a very popular course on the biological aspects of global change linking students at six sites (with students at the hub site in addition to the five major Colleges) and with lecturers from each site presenting a wealth of visual material.

Non-teaching applications of the network include a variety of academic meetings, library activities with demonstrations of database access and practical demonstrations such as book preservation, consortium meetings of research groups, and the introduction of academic staff to presentation skills.

Outreach

In the introduction to this chapter LIVE-NET was described as an 'island' network. This term is used to describe an enabling technology which is bounded to a small geographical region for reasons of range, availability or, perhaps, cost. An example of an island network in the digital signalling domain would be an Ethernet installed as a local area network (LAN) on a campus, or within a plant, which cannot be extended beyond the boundary of the plant because of node range limitations and/or because the PTO (Public Telecommunications Operator) does not provide a homogeneous service using the same technology. The PTO provides the wide area network (WAN) as an interconnect for island networks, using technologies that are not range dependent. Examples of WANs are X.25 packet switched networks carried over leased circuits up to 2 Mbit/s and, in the future, ATM (Asynchronous Transfer Mode) networks operating at 34 Mbit/s and above.

LIVE-NET extends well beyond the range of a local area network and can be described more appropriately as a metropolitan area network (MAN). The boundary limitations of LIVE-NET were initially determined by the maximum range of transmission of analogue television signals over fibre-optic cables within an acceptable degradation of signal quality. The longest installed link length to a site is 40 km which was just achievable within the flux budget available. More recent developments in fibre-optic transmissions systems would extend this range significantly but the boundary is now set by link installation cost.

Given that we wish to communicate beyond the range of the installed LIVE-NET network, what options are available to do this? Figure 13.2 shows the current outreach capability of LIVE-NET into both analogue and digital domains.

Cable

Perhaps the most obvious outreach is in to local community cable television networks (CATV) which employ a similar technology. Our interest here has been both to investigate the possibility of network extensions using the circuits of a cable operator and also to provide programming for community and educational channels. Although there are plans to inter-connect cable franchises in London there is no possibility of wider national or international access via this route.

Satellite

Following the launch of the European Space Agency (ESA) Olympus satellite in 1989, LIVE-NET was extended to an Imperial College site at Silwood Park for the deployment of a transportable earth station uplinking to the European DBS beam of the satellite. Over 70 programmes with interactive return links via a telephone bridge were transmitted before the loss of Olympus in 1991. For an example see Cain *et al* (1992).

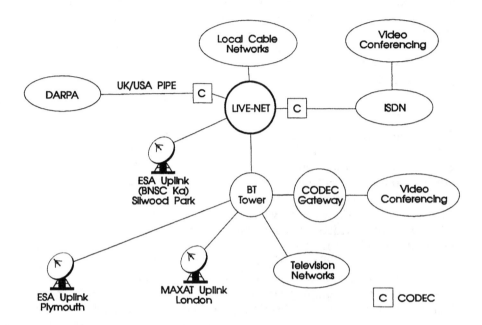

Figure 13.2   *LIVE-NET Wide Area Outreach*

LIVE-NET was further extended to the BT Tower in London for access to a range of services including other satellite uplinks in London and Plymouth. All such satellite broadcasts were one way video transmissions with audio return. A Ka band uplink with two way video capability to Canada was installed at Silwood Park for an experimental period.

Codec gateway

The link to the BT Tower can also be used to reach digital videoconferencing services offered by BT through a codec gateway in London. This gateway is equipped with codecs from different manufacturers with a range of standard and proprietary coding algorithms. LIVE-NET inter-links directly to the analogue side of the codec with connectivity to BT's national and international videoconferencing network.

DARPA network

One alternative approach to the use of dedicated circuits for videoconferencing is demonstrated by the experimental packet video link between UCL and the US DARPA network (Kirstein and Beckwith, 1991). By using a packet protocol the videoconference can be carried simultaneously with other network traffic (for example electronic mail) over the same logical link. The disadvantage of this approach is that the flow of packets for a time critical application such as videoconferencing cannot be guaranteed, resulting in periods of frozen images.

ISDN

Until the availability of ISDN the only digital circuit options were leased lines operating from 64 kbit/s to 2 Mbit/s. Although well suited for computer-to-computer networking, depending on the load capacity required, the usage profile of video networks is very different, making leased dedicated circuits of this type unattractive. A compromise can be reached in networks where two point-to-point connected sites require both a data and a video interconnection. Here a substantial proportion of the digital bandwidth of (say) a 2 Mbit/s circuit can be assigned to computer data traffic leaving 0.5 Mbit/s or less for video communication.

Within LIVE-NET the usage profile is naturally determined by the 10 week terms of the academic year and timetabling within the working day. This results in a network utilisation of around 30% of the total network capacity taken over a year with typically 30 hours per week during term time. Where courses could also involve students at sites in the wider area, the required link time would be a fraction of this, making access-on-demand connections to these sites through ISDN very much more cost effective and flexible than leased circuits. A similar pattern of usage would be expected in other organisations concerned with interactive education and training applications and with event based activities.

In the UK the initial offering of ISDN from BT was 2 Mbit/s ISDN-30 primary rate (30 x 64 kbit/s) for the connection mainly of company PABXs. This service was followed later by the availability of Basic Rate ISDN circuits at 2 x 64 kbit/s. For LIVE-NET outreach through ISDN we had installed, at the LIVE-NET hub in Senate House, four BT ISDN-2 circuits with the objective of providing a versatile set of connections. For example a low cost BT VC7000 videoconferencing terminal can be connected to one Basic Rate circuit for 128 kbit/s operation or, alternatively, three of the Basic Rate circuits can be aggregated through an ASCEND inverse multiplexer to provide a 384 kbit/s link for a more sophisticated codec. The Basic Rate circuits are also available to be used individually for terminal access to LIVE-NET resources such as the Image Server. Applications using these devices are described in later sections.

The essential advantage of ISDN over other outreach options is in its connectivity both within the UK and internationally. Using the ISDN-2 service LIVE-NET can immediately interconnect with Universities in (for example) Europe, Australia and the Far East for visual conferences and meetings without the need for any prior booking arrangements with satellite agencies. The LIVE-NET ISDN-2 circuits and codecs provide a gateway function for any of the LIVE-NET locations that wish to gain access to this facility.

Image Server

The concept of the LIVE-NET Image Server was first developed as part of a DELTA Exploratory Phase programme CAPTIVE (Collaborative Authoring, Production and Transmission of Interactive Video for Education) and subsequently enhanced in the DELTA '92 programme MTS (Multimedia Teleschool). The Image Server comprises a Sony analogue WORM (Write Once Read Many) optical disk system and a control program initially developed for MS-DOS and later for a multi-user Unix environment. The optical disk is a cartridge with a capacity of 36,000 still images or 24 minutes of full motion video (or any combination of still and moving images). Frame by frame recording is at near broadcast quality and access to any part of the disk is within 0.5 second with full motion playback.

The objective of CAPTIVE was to demonstrate that video sequences could be catalogued and re-used for new productions. Short video clip excerpts from produced linear programmes were recorded on to a cartridge disk with associated retrieval information stored in a database. For MTS the concept was widened to include visual material for pilot courses. Within LIVE-NET the system has been used by lecturers to pre-store all of the pictures and diagrams required for a series of lectures.

The Image Server is linked into the LIVE-NET hub switch and has been used as an integral part of satellite transmissions through Silwood Park to the Olympus satellite, and to the MAXAT uplink in London to Eutelsat II-F3.

ISDN access

Of particular interest here is the inter-connection of the Image Server to ISDN via LIVE-NET for remote retrieval of stored images, image loading and browsing (Duval *et al*, 1993). As part of an international MTS multimedia workshop organised by France Telecom DFTP (Direction de la Formation Professionelle) in Montpellier, the Image Server was used to record key presentations being delivered by satellite under the VIF (Videocom Interactive France) system. (VIF transmits from studios in

Montpellier at broadcast quality by satellite with interactive videoconferencing returns from participating sites by leased lines and ISDN). These presentations were sectioned and catalogued into a database held on the LIVE-NET Unix control computer. During a later part of the workshop the Image Server was linked back to Montpellier via the ASCEND inverse multiplexer at 256 kbit/s (2 x ISDN-2 circuits) for interactive playback of any section to workshop participants under the direct online control of the presenter at DFTP.

ISDN is also being planned for the uploading of visual material into the Image Server. Presently this is being achieved through the use of a proprietary Sony Digital Information Handler (DIH) over PSTN for a course in Electronic Engineering from the TVEI Centre Gwynedd in North Wales. This is being undertaken as a pilot course demonstration of the Image Server within MTS. Using ISDN from a PC platform and employing standardised protocols for image compression and transfer will make the uploading capability more widely available. It is also proposed to use the same PC platform for students and teachers to be able to browse images held in the Image Server between satellite transmissions.

Multi-site interaction

LIVE-NET

The configuration of LIVE-NET for a fully interconnected interactive session with one site per video channel has been described previously. Using a multiplexing technique and specially developed audio mixing system, up to eight sites can be interconnected with simultaneous vision and sound using only three receive channels at each site (Figure 13.3 on next page).

Two multiplexers each take four non-synchronised video transmit channels and form composite 2 x 2 quadrant images which are fed back to the hub switch for distribution to each site over two receive channels. The third channel is used for selection of one of the quadrants to be displayed as a full screen main picture. This may be used to display the site which is speaking at any time or to display a graphic which is being transmitted from one site and being discussed by sites in general.

The control of the main picture may be under the direction of one site (chair control) or by each site separately (user control). Selection is effected through a terminal linked back to the LIVE-NET switching computer. Picture switching occurs within a faction of a second. Similar terminal control techniques have been developed for real-time editing applications (for example to solve the problem of producing a video tape recording from a multi-site session without the need for multiple recorders) and for image selection for connected sites that do not have full channel capability. The latter application is described in the next section.

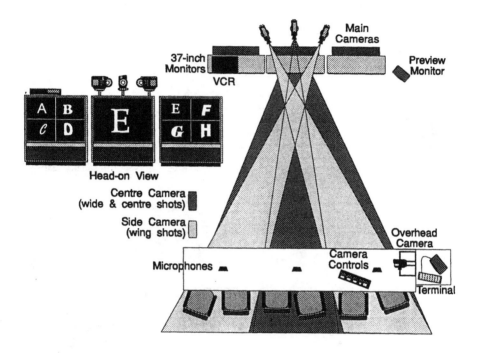

Figure 13.3 *LIVE-NET room showing multiplexed session between eight sites*

LIVE-NET and ISDN

The first academic use of the integration of ISDN into LIVE-NET occurred with the holding of a joint postgraduate course in HCI (Human Computer Interface) between Imperial College, University College and Birkbeck College, London and Magee College, Ulster. The course content and ISDN issues are described in detail in another chapter (Chapter 5). Here we will discuss the interconnection of the videoconferencing terminal and picture control.

The first models of the BT VC7000 (BT VC7000, 1992) desktop videoconferencing terminal became available for evaluation in the UK at the beginning of 1993 (BT, 1992). Developed by Tandberg in Norway the terminal is a completely self-contained system comprising a monitor, video camera, codec, control keypad and ISDN terminal adaptor. Although designed primarily to be a used as a desktop system the VC7000 can be connected directly to external video input/output devices and audio equipment. This makes the terminal ideal for interconnection to the LIVE-NET hub switch for access from any LIVE-NET site or a group of sites (Figure 13.4).

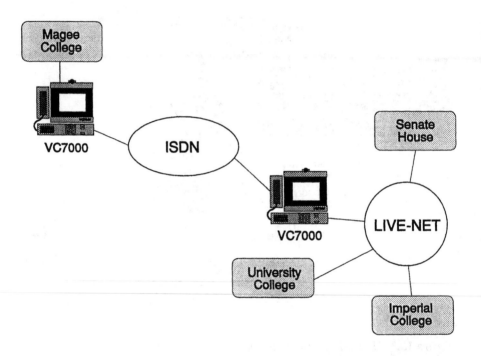

Figure 13.4   *Integration of LIVE-NET and ISDN for a four site session*

In London three sites were participating in the HCI course from locations on LIVE-NET, with Imperial College and University College the main lecturing and student sites with Birkbeck students also present at the hub site (Senate House). The VC7000 was set to automatic connect mode which enabled Magee College to dial in and connect immediately into the session without any technical intervention. The receive channels from the three London sites and the received image from Magee were combined using a picture multiplexer. This quadrant image, or alternatively a full screen image from any one of the three London sites, could be switched back through the single channel of the VC7000 to Magee depending upon the context of the session. Students at Magee would receive a full screen picture while a lecturer was speaking from one of the London sites and a multiplexed quadrant during discussions. Students in London saw each site, including Magee College, on a separate channel. A combined audio feed of the three London sites was transmitted to Magee throughout the session.

For this first course, control of picture switching was retained in London to simplify session management during the experiment. However, as in the case of the Image Server, this is a networked function and future courses can have control of switching remotely either through the wide area digital network or, possibly, via the digital channel supported by the VC7000.

ISDN hub

The VC7000 is a relatively low cost complete videoconferencing terminal in comparison to high performance fully functioned codecs. This has been made possible through recent advances in codec design and manufacture which will result in single board codecs becoming available within a year. The significance of these developments is that it becomes possible to consider videoconferencing networks presently operating picture switching via an MCU migrating to visual communications networks with multiple channels and full telepresence, i.e. the LIVE-NET model.

The key to this future is the functionality and implementation of a network hub. In order to achieve telepresence, the transmitted images from each site must be available at the hub for mixing and redistribution. An MCU basically switches the received data streams and does not contain any codec functionality to recover images from the data stream. With low cost codecs becoming available, a hub can be implemented as an ISDN network service with a codec on each callable circuit. The analogue outputs from the codecs can be routed to a matrix switch and multiplexers to achieve a similar functionality to LIVE-NET.

Thus in a multi-way conference, all participants can be fully viewable at each location. Further the same kind of real-time picture switching control can be implemented using keyboard control via part of the ISDN bandwidth. An immediate route to the implementation of a prototype would be to place four VC7000s on the installed ISDN links to LIVE-NET and use the existing switching and mixing equipment to provide the ISDN hub functionality.

Future directions

So far we have discussed the way in which ISDN has been, and might further be, integrated into the LIVE-NET broadband network. We will look finally at options for a successor network given that analogue broadband technology will give way inevitably to digital based network architectures. An issue discussed previously is that of the benefit of on-demand networks compared with fixed circuits. Certainly changes in funding and patterns of usage suggest that networks which can offer a low connect cost, flexible bandwidth, and charging only for connect time used, will be more marketable than fixed circuits for educational video applications.

In the UK the networks offered by BT and Mercury for on-demand access (ISDN and Switchband respectively) can meet certain of these requirements although tariffs are mainly targeted at acceptable levels for business users rather than educational users.

Given that the network can be provided economically, will the users of an existing broadband network with full bandwidth television channels, and to whom videoconferencing is not a novelty, accept the quality of codec based network operating at conventional data rates (between 128 kbit/s and 2 Mbit/s)? This issue is presently being evaluated with side-by-side comparisons between full bandwidth and codec images. An interesting observation is that users expect picture resolution, in addition to motion rendition, to increase with bandwidth. With the H.261 standard this is not the case although proprietary algorithms exist that do improve resolution at higher rates.

Another issue is whether room based systems will be phased out in favour of the integration of videoconferencing into networked personal workstations. This is likely to follow current trends in higher education away from didactic teaching to resource based and flexible learning. With these considerations the quest for a successor network to carry forward the aims of LIVE-NET will be both a challenge and an opportunity.

References

BT. VC 7000 Videoconferencing System (1992). BT data sheet PHME 11685/8/92.

Cain, G. D, Morling, R. C. S. and Yardim, A. (1992). Trans-European experiments in DSP education. *IEEE Signal Processing Magazine*, 1992, pp 20-24.

Duval, E. *et al* (1993). An information and telecommunications infrastructure for flexible distance learning. *Proc Telepresence '93*, pp 58-66.

Kirstein, P. T. and Beckwith, R. C. (1991). Experiences with the University of London interactive video education network. *Electronics and Communications Engineering J*, **3**, 1, 4-12.

Powter, E. J. and Beckwith, R. C. (1988). LIVE-NET – an interactive switched video network for the University of London. *Int Journal of Digital and Analog Cabled Systems*, **1**, 1, 19-28.

CONCLUSIONS

The previous chapters have made the point that videoconferencing is one of the main applications of ISDN in education and training. Although it is not the only one, it is so eye-catching (literally) that it can distract attention from other, humbler applications which may perhaps in the long term be nearly as useful.

Videoconferencing

Advantages

Videoconferencing can and has been used successfully in a wide number of applications. In education and training, it has several advantages over rival media:

- Videoconferencing is meeting current demands for more open and flexible learning. It is particularly appropriate for professional updating and other continuing education and training areas. It also extends course choice to smaller or isolated campuses.
- The instructors accept it because it does not seem to impinge too much on their pre-existing presentational approach. (They accept that it requires higher-quality visual aids, but the quality of visuals is rising anyway with the increased use of desktop publishing techniques. Perhaps more reluctantly, they accept that there are human presentation skills that must be acquired too.)
- In some ways a videoconference can be " better than being there" (such as gaining a better view of certain procedures or equipment).
- The administrators of education and training systems understand the videoconferencing methodology because it maintains the concept of contact hours, thus apparently making the amount of teaching easy to measure. (We pass over the complex issues of the amount of learning that has taken place.)

Disadvantages

The following three points summarise an extensive debate on the pedagogical and technological disadvantages:

- There is a general disadvantage that videoconferencing, if used merely for video lectures with no feedback, will foster a 'transmissive' model of education, currently out of favour with educational thinkers (though not with administrators who tend to like the transmissive model's economies of scale).
- In the candid classroom situation, where students are viewing the lecture locally as well as over a network, there is a problem of balancing the interaction requests of the local students with those of the remote ones. The solution is not to do away with the local students – instructors value their presence.
- The most operationally convenient, and the cheapest, approach to network support for videoconferencing is to use Basic Rate ISDN, with its two B-channels, giving an aggregate data rate of 128 kbit/s. Yet the quality of video at this rate is only adequate, especially when using standard codecs. Balancing the virtues of standardisation (better connectivity, wider range of suppliers, future-proofing, etc) against the problems (poorer functionality, possibly higher cost, later delivery, etc) is always a difficult activity. Some applications such as medical videoconferencing are likely to continue to need higher bit rates than Basic Rate ISDN. (Examples include six aggregated ISDN B-channels giving 384 kbit/s.)

Action to overcome the problems

1 Better training of instructors is needed:
   - both so that they can use videoconferencing more effectively when it is appropriate to use videoconferencing
   - and so that they can select media better, ensuring that they *know* when it is appropriate to use videoconferencing.
2 Education and training managers need a better understanding of the roles of educational media and their cost-benefits, and a clearer picture of the kind of infrastructure investments required.
3 Further technical development of codecs is needed so that 128 kbit/s videoconferencing offers higher quality video. It is important to co-ordinate such work with the developments going on in digital television and CD-ROM.
4 Work must continue on standardisation of the 'peripheral' (but increasingly vital) functions of videoconferencing systems such as camera control and slide scanners.

ISDN applications

The early adopters of ISDN in education and training are finding successful uses. However, there are some limitations:

Speed: a number of articles in the book have touched on the fact that the Basic Rate ISDN channel configuration is not really fast enough at the present level of technology. Videoconferencing needs expensive and/or proprietary codecs and/or channel inverse multiplexing techniques. Another problem is that large image files still take a long time to transmit. Primary Rate ISDN is not the answer, for technological reasons and because it is far too expensive.

Coverage: in many countries ISDN is still limited to major cities, yet many potential educational requirements for ISDN are oriented to more remote areas since these are precisely the areas that conventional lecture-based approaches find hard to support.

Ease of use: although the vendors of Basic Rate ISDN videoconferencing have made great progress in making ISDN easy to use, in other applications there are often complex configuration procedures to undergo.

Lack of awareness:

a There is a lack of awareness of the potential of ISDN for the small business and domestic market. Without the economies of scale created by rapid growth in these sectors, ISDN equipment is likely to remain too expensive for mass use in the education sector. The PTTs (as the main vendors of ISDN services) need to do more, both to promote the benefits of ISDN and to understand the real needs of users.

b In the education and training sector, there is a lack of awareness of the educational potential of ISDN-based educational systems (such as videoconferencing and audiographics). This is part of a larger problem of lack of awareness of the potential of telematic networks, and more generally of information technology, and more generally still, of distance education techniques.

Future prospects

The concept of ISDN goes back to the 1970s and the first standards were ratified in 1984 (in the CCITT Red Book). Yet many developed countries do not have ISDN yet and in those that do, it is still little used. Thus it is fair to say that ISDN has not been an overnight success. Nevertheless, at last (and not just in Europe) ISDN has got off to a good start. If nothing else, we hope that our book has convinced you that, in both a technological and a pedagogical sense, ISDN works – and education and training applications work over ISDN.

Glossary

Paul Bacsich

Introduction

This glossary does not claim to be a comprehensive list of all the terms relevant to ISDN. There are several well-known dictionaries of Information Technology that will give you that.

Instead, it aims to contain all the ISDN terms that *non-experts* are likely to require – for example, in order to plan services and discuss specific systems and services with vendors. It also includes (more unusually) a survivor's guide to the main *educational* terms relevant to (our perception of) the use of ISDN in education and training.

So next time that your videoconferencing vendor looks askance at you when you do not understand the nuances of H.261, ask sweetly whether their system is really appropriate for a third-generation distance education system or is still stuck in a transmissive paradigm...

A

A/D conversion
: Conversion of information from analogue form (such as the loudness of a sound) to digital (so that it can be represented in a computer).

Accunet
: A switched 56 kbit/s service offered by AT&T in the United States which is in many ways a precursor of ISDN.

analogue
: Information without any discernible steps, which can vary in a continuous fashion (e.g. loudness of sound)

ASCII
: American Standard Code for Information Interchange. A method of encoding the letters, numbers and special characters used in computing as

integers between 0 and 127. Thus for example the letter A is coded as 65.

asynchronous

This term is well-known in data transmission but is used differently in educational circles.

i) In data transmission, asynchronous transmission is a method of transmitting data in which each character of data is sent separately, at unpredictable intervals. The *V.24* serial interface (as used in microcomputers and modems) supports asynchronous transmission.

ii) In education, asynchronous interaction is where the two parties are not (or need not be) simultaneously present. Thus computer conferencing is regarded as an asynchronous medium, whereas videoconferencing is *synchronous*.

ATM

Asynchronous Transfer Mode. A key enabling technology for Broadband ISDN using a packet switching approach with fixed-length cells.

audio bridge

A method of connecting a small number of telephone lines in order to provide an audio conference. Audio bridges over the PSTN have to cope with different audio levels and background noise conditions. It is hoped that on the ISDN, these problems will disappear.

audiographics

A small range of technologies which combine a live voice link with a shared screen for computer graphics, real-time drawing or pre-prepared material. It is a technology which has never become truly widespread and is currently in transition, partly due to ISDN.

audiotex

A system which allows users with a touch-tone phone to access and interact with a database of audio messages. In recent years audiotex has become popular, due largely to the reduction in cost of systems (which can now be implemented on microcomputers) and the integration with voice mail and PBXs.

B

B-channel | The main type of component channel of ISDN services, used for carrying data or voice. (B stands for "bearer".) *Basic Rate ISDN* consists of two B-channels and one D-channel. Primary Rate ISDN consists of 30 B-channels plus one D-channel (or 23 B plus one D in North America). A B-channel has a bit rate of 64 kbit/s.

B-ISDN | Broadband ISDN. The proposed successor to *ISDN* that will operate at *broadband* speeds, even to the home (but not over the existing phone cabling).

bandwidth | The range of frequencies required to transmit a signal. For example, voice over the telephone network requires a bandwidth of 3 kHz while uncompressed video requires a bandwidth of 6 MHz.

Basic Rate ISDN | This is a version of ISDN which consists of two 64 kbit/s channels (B channels) for speech or data, plus a 16 kbit/s channel (D channel) used for signalling and control purposes. The aggregate data rate is thus 2 x 64 + 16 = 144 kbit/s. Basic Rate ISDN is often referred to as 2B+D.

baud rate | The number of changes of state per second of an item of information. Must be carefully distinguished from bit rate, but many people confuse the two, probably because many lower-speed modems have the bit rate and baud rate identical.

behaviourist | A pejorative phrase associated with second-generation distance education. A behaviourist approach to teaching is often associated with use (or over-use) of Computer Based Training.

bit | A BInary digiT. The 'quantum' of information., A bit has only two values, On or Off (conventionally coded as 1 and 0, although people disagree as to whether 1 means On or Off).

bit rate | The number of bits per second in a bit stream. For example, the bit rate of widely used modems is in the range 300 to 14400 bits per second. The bit rate of ISDN is 64000 bits per second.

bit stream

A sequence of bits, usually regarded as potentially endless and occurring at regular intervals.

BITNET

A wide-area computer network linking many universities in North America.

Blue Book

The 1988 edition of the CCITT reports which, among many other things, control the "I-series" standards used for ISDN. (See Volume 3, Fascicles 7-9.)

bridge

A device which connects telecommunications or data communications channels. There are two distinct areas of meaning:

i)   In teleconferencing, a bridge connects three or more channels. Thus *audio bridge*, videoconferencing bridge, etc. Co-Learn is building an audiographic bridge (see Chapter 11).

ii)   In data networking, a bridge connects two local area networks so that they appear to be one network.

broadband

This is a hard to define term. The real meaning is "faster than commonly occurring networks", and thus the meaning depends on the year in question. Many people regard the boundary between broadband and narrowband as occurring at 2 Mbit/s. Whether this makes 2 Mbit/s itself broadband or narrowband is not clear; and 2 Mbit/s is now not so fast, even for wide area networks. A better approach is to regard the boundary as occurring somewhere around 20 Mbit/s – under this classification all commonly occurring local area networks (Ethernet and Token Ring) are narrowband; only FDDI (at 100 Mbit/s) would be broadband.

broadband ISDN

A version of ISDN with all the ISDN advantages (such as dial-up) but working at broadband speeds, conventionally agreed to be above 2 Mbit/s. Should be distinguished from Primary Rate ISDN which works at 2 Mbit/s but is 'only' a bundle of 64 kbit/s ISDN circuits, not a fully integrated service. The two main proposed Broadband ISDN rates are 150 Mbit/s and 600 Mbit/s.

broadband ISDN applications

> PTTs are notoriously bad at predicting the applications for their new services. The following list should be taken as general guidance only. They are essentially 'ISDN applications but more so'. The last three are particularly vulnerable to advances in video compression and the last two to the state of the economy.
>
> i)   LAN to LAN interconnection, especially for backup, occasional extra capacity and disaster recovery.
>
> ii) Transfer of very large image files such as are found in medical and newspaper applications.
>
> iii) Sophisticated multimedia conferencing and group work generally.
>
> iv) HDTV to the home.
>
> v)   Access to video databases (for entertainment, home shopping, etc).

BT
: The main UK telephone operator previously known as British Telecom.

byte
: A group of 8 bits. The internal memory of most microcomputers is structured in bytes. It is convenient to think of a byte as carrying one character of text.

C

CBT
: Computer Based Training.

CCITT
: Comité Consultatif International Télégraphique et Téléphonique (International Consultative Committee for Telegraphy and Telephony). An agency of the International Telecommunications Union (*ITU*) dealing with standardisation and regulation of the public telecommunications sector, including modems, ISDN and videoconferencing. In several areas its activity overlaps with *ISO* but the two agencies endeavour on the whole successfully to avoid conflicting standards.

CD-ROM
: Compact Disc Read Only Memory. The adaptation of compact disc technology (as used for hi-fi audio) to

store data. Up to 600 Mb of data can be stored on one CD-ROM.

CD-ROM is often seen as a rival to ISDN in the application area of distribution of information – instead of providing a central database accessible by ISDN one 'merely' duplicates the information on CD-ROMs and sends them out by post. However, it is more realistic to say that the application areas overlap but are distinct – in particular CD-ROM cannot be used for information that changes rapidly (even as rapid as weekly) while ISDN cannot be used for access to complex multi-media material (such as naturalistic motion video) unless very expensive codecs are used (note that the transfer rate from CD-ROM is much higher than ISDN speeds).

CEPT

The European grouping of PTTs who undertake in Europe functions similar to that of CCITT.

character set

A set of letters, numbers and punctuation symbol used in text to be processed by a computer. The main character set in common usage is the *ASCII* character set – this has 128 characters (thus requiring 7 bits of information). Other common character sets (such as those used on microcomputers and for European telematic services) have 256 characters and thus require 8 bits (one byte) of information

circuit switching

A methodology of connecting two systems which relies on creating a communications path between them for the duration of the session, probably through intermediate exchanges. The PSTN and ISDN use circuit switching. One may wonder how there could be any other way of doing things; but see *packet switching*.

CLI

Compression Labs International. One of the leading vendors of videoconferencing systems.

CMC

Computer Mediated Communications. A phrase normally used to mean *computer conferencing, electronic mail* and access to remote data bases.

codec

COder-DECoder. A term (constructed analogously to modem = modulator-demodulator) to mean a device which translates between analogue video and a compressed digital format. The bit rate of the digital signal from a codec is normally in the range 64 kbit/s to 2 Mbit/s, depending on the sophistication of the codec and the degree of degradation of the video signal which is acceptable in the context in which the codec is being used.

collaborative learning
A mode of learning in which a group of students work together on a topic or project.

compressed audio The use of compression techniques to transmit audio over low-speed data circuits. In an ISDN context, audio compression is used as part of the compressed video system. ISDN-quality voice (at 64 kbit/s) can be compressed to about 16 kbit/s without significant loss of quality. There are compression techniques (for example used in military and satellite systems) which can produce recognisable voice at 9.6 kbit/s and below, but they do not have acceptable quality and performance for use in videoconferencing.

compressed video The use of compression techniques to transmit colour motion video (plus sound) over digital circuits such as ISDN. Older techniques – such as *H.120* – could only compress video into 2 Mbit/s or perhaps 384 kbit/s. Newer techniques – such as *H.261* – can compress video into a 128 kbit/s channel (two ISDN circuits), or 112 kbit/s in the US.

compression A range of techniques which reduce the amount of data required to carry a specific amount of information. There are general compression techniques which apply to any data, but far better results are obtained when the compression techniques use information about the general behaviour of the information, such as text, audio, image or video.

computer based training
The general meaning is the use of computers to assist the training process. However, the term is normally used in a more restrictive sense to mean the use of software on a computer to teach, help and/or assess coursework. The basic idea is that computer based training can be used to replace to some extent the use of human trainers.

computer conferencing

A development of electronic mail designed for effective support of many-to-many communication. Conferencing software includes features specifically designed to help in the organisation, structuring and retrieval of messages. Messages can be linked to each other (e.g. as 'comments'), organised in different 'branches' or 'topics' of a conference. Search commands can rapidly identify messages with particular keywords in their titles or in the body of the text.

Special commands are available to the person responsible for a conference (the 'moderator' or 'organiser') which can help in defining the membership of the conference, in keeping the discussion on track, and in scheduling the opening and closing of discussion topics.

Computer Supported Collaborative Learning

The use of computer systems (such as computer conferencing) to facilitate collaborative learning.

Computer Supported Co-operative Working

The use of computer systems to facilitate co-operative working. Systems to support this include electronic mail, computer conferencing, group scheduling systems, databases and shared desktop systems.

conferencing

A generic term used to cover various types of system which link people together. The main variants are videoconferencing, audioconferencing, audiographic conferencing and computer conferencing. All except the last link people together 'synchronously', that is, the people are present simultaneously even if separated in space.

CSCL   Computer Supported Collaborative Learning

CSCW   Computer Supported Co-operative Working.

D

D-channel

The control/signalling channel in ISDN. (D stands for "delta".) In *Basic Rate ISDN*, which consists of two B-channels and one D-channel, the D-channel has a bit rate of 16 kbit/s. In *Primary Rate ISDN*, which consists of 30 B plus one D, the D-channel has a bit rate of 64 kbit/s.

D-channel uses   The CCITT have proposed that the D-channel could be used for access to public packet switching networks. Calculations indicate that even under conditions of heavy signalling use, over 95% of D-channel capacity is available for user services. The 16 kbit/s rate of the D-channel would make a very elegant enhancement of the 2400 or at best 9600 bit/s dial-up access currently offered for packet switching services. However, PTTs have been slow to offer packet switching over the D-channel. It is thought that worries over the appropriate tariffing of the D-channel have delayed matters, since PTTs do not want to jeopardise their revenue from PSTN dial-up access to packet switching services

D/A conversion   Conversion of information from digital form (such as in a computer) to analogue form (such as sound).

D2-MAC   A standard for satellite TV signals, often seen as a halfway house to High Definition Television. It is not widely used and looks likely to be superseded by digital technology.

data compression   The application of compression techniques to general data. Often (but not always) this data is text. The CCITT standard data compression technique for low-speed data transmission is V.42 bis, which is gradually replacing older and more proprietary techniques. V.42 bis will yield up to 4:1 compression.

DBS   Direct Broadcast Satellite.

deferred-time   Several authors (see for example Chapter 12) use deferred-time as a synonym for *asynchronous*. You will also find the horrible phrase "non-real-time" in some literature. From Chapter 12 by Zorkoczy *et al*: "The possibility to communicate with co-learners and tutors in deferred time adds a new dimension to collaborative educational settings. If the learner is allowed to answer a question later, that is, not on the spot as in traditional classroom scenarios but after a period of time, a new type of learning process will evolve with benefits for some learning situations. The immediacy of the group situation will be diminished in favour of the opportunity to work on responses with more contemplation and depth."

DELTA   Development of European Learning through Technological Advance. A European Commission DG XIII funded programme for research and development in Telematic Systems for Flexible and Distance Learning.

desktop conferencing
                     The use of desktop microcomputers for videoconferencing, shared screen and audiographics.

digital   Information represented in bit form.

Direct Broadcast Satellite
                     A type of satellite which transmits video direct to simple receiving dishes on domestic or institutional premises. DBS satellites are to some extent a rival to ISDN for transmission of video – if there is a requirement to transmit the same video to many sites without the need for a video return channel then DBS will be cheaper than using many ISDN codecs. Conversely if a video programme needs distribution to only a few sites then ISDN videoconferencing will be cheaper than DBS.

distance education   Whole books have been written on the meaning of this term, and we do not intend to write another. Thus the following four brief thoughts are open to challenge.

i)   Distance education is the use of techniques to allow the education of students elsewhere than on the campus or in the classroom.

ii)   The techniques do not need to use advanced technology – much distance education uses print and videocassette only.

iii) There is no rigid dividing line between distance education and 'conventional' methods – distance education techniques can be applied to students enrolled on campus-based institutions, and lecturing and tutorial techniques can be applied to students enrolled in distance teaching institutions such as the UK Open University.

iv) ISDN allows the relatively easy adaptation of conventional techniques (lectures and tutorials) to the distance situation.

downlink   A satellite earth station which receives signals (such as television) from a satellite. The opposite of *uplink*. See also *TVRO*.

DTC   Desk Top Conferencing.

E

EARN   European Academic Research Network. A computer network for universities in Europe originally set up by IBM.

echo   On a telecommunications channel, echo is the reflection of a portion of the signal energy from the remote end back to the transmitter. In audio terms, this is perceived as a faint and delayed version of the speakers voice. Control of echo is a major problem in telecommunications. It affects modems, audio bridges, loudspeaking telephones and even videoconferencing over ISDN (because of the existence of microphones and loudspeakers).

electronic mail   The electronic transmission and reception of information in *asynchronous* mode – that is, without the sender and recipient needing to be simultaneously present. Originally electronic mail was text-based, and most systems on wide area networks are still restricted to text, but on local area networks most modern electronic mail systems can transmit arbitrary data including image, audio and even video messages (if the bandwidth is sufficient).

Computer conferencing is a more sophisticated form of electronic mail.

electronic whiteboard
    Originally, a device based on a whiteboard which transmitted the pen movements to a pen moving over another whiteboard at the remote end. Now often applied to screen-based systems with similar functionality.

email   An abbreviation (but not used in this book) for *electronic mail*.

Ethernet
: A popular local area networking technology operating at 10 Mbit/s over coaxial or twisted pair cable, and increasingly over optical fibre. Despite operating at an apparently high speed, Ethernet is not really suitable for the transmission of real-time signals such as speech or compressed video.

ETSI
: The European Telecommunications Standards Institute. An organisation set up only recently (in 1988) by the European PTTs with the support of the European Commission which is taking over much of the standardisation work previously done by CEPT and other bodies. ETSI is currently looking at harmonisation of ISDN standards across Europe.

Euro-ISDN
: A harmonised version of ISDN which will eventually be operable in all European countries. Those countries who implemented ISDN the earliest will have to make some changes in their systems.

EUROSTEP
: EUROpean association of Satellites in Training and Educational Programmes. An association of organisations interested in developing and piloting satellite TV services for education and training. EUROSTEP transmits educational television on one of the EUTELSAT satellites.

EUTELSAT
: EUropean TELecommunications SATellite organisation. Owned by the European PTTs, they run the main European communications satellites.

EWB
: Electronic WhiteBoard.

experiential learning
: A range of ideas connected with the concepts of 'learning by doing' and 'trial and error'.

F

face-to-face
: The traditional method of teaching, in a lecture theatre or seminar room. The most obvious applications of ISDN in education and training are those which substitute the new medium for the old – thus videoconferencing seems to allow the lecturer to be at a distance, not face-to-face, without changing the other parameters.

There is often thought to be an ideological gulf between proponents of distance education and proponents of face-to-face teaching. In several chapters you will see that this gulf is both bridgeable and being bridged.

facsimile

A technique for transmitting text and black-and-white pictures over the telephone network. Facsimile currently operates at 9600 bit/s over the PSTN but there is a standard – *Group 4* – for facsimile over ISDN which may become popular.

fps

Frames per second.

frames per second

A European standard (PAL) television signal displays 25 frames per second and a US standard (NTSC) signal displays 30 frames per second. Various kinds of compressed or simplified video (such as QuickTime) often use less frames per second but the results are not naturalistic.

full duplex

A telecommunications channel which allows 'conversation' to take place in both directions at once. The typical data communications channel (the *V.24* interface) between a microcomputer and a modem is full duplex.

full motion video

A video signal before it has been processed by video compression. However, the term is often used to denote a *digital* signal, so that the precise meaning does depend on the sampling used (as well as on the TV standard used). For example, a PAL signal transmits 25 frames per second. A 640 by 480 display with 24 bits of colour information per pixel would generate 184,320,000 bits per second of information – 184 Mbit/s! Perhaps now one sees why compression is needed.

G

GPT

GEC Plessey Telecommunications. A well-known manufacturer of codecs.

graphical user interface

A windowing user interface similar to that found on the Apple Macintosh microcomputer or on MS Windows systems on PCs.

Group 3   The current mass-market standard for facsimile over PSTN. This operates with 9600 bit/s modems, which does not seem very fast, but complex data compression techniques are used to reduce the average transmission time of an A4 page to about 30 seconds.

Group 4   A new CCITT standard for facsimile oriented to use over the ISDN. This offers resolution, grey level, error correction and data compression facilities all better than Group 3. It is designed to allow transmission of a typical A4 page in about 1.5 seconds over one ISDN channel.

Group 4 facsimile was seen by the PTTs as one of the winning applications for ISDN. However, early Group 4 machines are extremely expensive, and *Group 3* technology is performing very well, especially now that it is integrated into microcomputers and electronic mail systems and supports output on plain paper (with good progress being made on automatic optical character recognition of incoming faxes). There are also worries that the Group 4 standard is not sufficiently integrated with the corporate world of laser printers and desktop publishing to be attractive in the office context.

Group 5   An emerging facsimile standard oriented to colour facsimile over Primary Rate ISDN.

groupware   A rather vague term covering a range of existing and emerging computer technologies used for computer supported co-operative working and computer supported collaborative learning. Examples include conferencing (audio, audiographic and video), electronic mail, computer conferencing, scheduling and diary management systems, shared desktop systems and multi-user editors.

H

H series   The series of CCITT Recommendations governing audiovisual services.

H.120   An older CCITT standard for compressed video. Note that H.120 is an older set of standards for 2 Mbit/s only and is not compatible with the newer H.261 set of standards.

H.261   A modern CCITT standard for compressed video, also known as the p x 64 standard. From Chapter 1 by Jacobs: "For ISDN, the CCITT recommends its H.261 video data-compression standard, introduced last year after considering all available compression methods; one of the principal requirements was compression and decompression on the fly. The H.261 standard incorporates numerous extra techniques such as a system of frame difference, in which each frame in a video sequence is encoded not in its entirety but only as the differences between it and the frame which preceded it. So, if two adjacent frames consist of a static background and a moving object, only the movement need be encoded, thus saving space."

H.320   A recent CCITT standard embracing the better known H.261 relating to video encoding algorithms. The standard is used in the VC7000 videoconferencing system. From Chapter 8 by Lange: " H.320 has deliberately frozen technology at a moment in time (1988) so that competing manufacturers can ensure inter-operability with each other. Major manufacturers have continued to developed proprietary algorithms that enhance or extend the standard to provide better audio and video quality and better feature sets. For example, proprietary algorithms allow acceptable video and wideband audio (7 kiloherz) while running at 128kbps, while the H.320 standard is acceptable at that speed only if narrowband audio (3.4 kiloherz, 16kbps) is used. Proprietary algorithms also contain special features, such as far-end camera control, which are valuable in teaching, but which are not included in the standard. . All of the major vendors (including PictureTel) sell an interoperating version of the H320 standard algorithm, though with some manufacturers, older equipment cannot be converted to run the standard without a major hardware change."

half duplex   A telecommunications channel which allows 'conversation' to take place in only one direction at a time. A telephone connection using loudspeaking telephones is half duplex because of the voice switching technology used. The typical style of interaction between a user and a mainframe computer (such as in videotex) is half duplex – the user types a command, the computer generates a response– yet the underlying serial communications channel is full duplex.

HDTV

High Definition Television. A concept of higher-definition wider-screen television being piloted in a number of countries.

hertz

The unit of measurement of frequency. See *bandwidth*.

hypertext

Another of these words which is difficult to define. Hypertext systems store text in 'chunks'. Users can navigate from one chunk to another by following links. The links are attached either to 'buttons' on the screen or to the words in the text chunks – or to both.

This glossary could be easily reconfigured as a hypertext – each glossary entry would be a chunk and each word in the glossary text would be a potential link to a related glossary entry.

hypermedia

A development of hypertext which incorporates multimedia (diagram, photographs, audio and video). Imagine if this glossary contained video clips describing some of the entries.

I

I series

The series of CCITT Recommendations governing data transmission over ISDN.

Integrated Services Digital Network

The CCITT definition is as follows: "A network, evolving from a telephony Integrated Digital Network (IDN), that provides end-to-end digital connectivity to support a wide range of services, including voice and non-voice services, to which users have access by a limited set of standard multi-purpose user network interfaces."

interactive

This term has several related meanings and a strong emotional content.

i) In telecommunications terms, an interactive channel is a full duplex channel – communication can take place in both directions at once.

ii) In media terms, an interactive medium is one with which one can react. Thus Computer Based Training and electronic mail are interactive media. So are interactive video disc and CD-ROM. Television and print are often said not to be interactive – but others argue that the students interact with these media 'in their heads'. Perhaps this is more true of videocassette than broadcast television because the student can stop, start and rewind the tape.

iii) By extension, an interactive medium is believed to be 'good' and 'modern' and a non-interactive one 'bad' and 'old-fashioned' – maybe even *transmissive*.

The following quotation (from Chapter 9 by Lafon) may put some of the above in context: "The students admitted that the interactive facilities were somewhat daunting: having to speak in public, with a camera searching for you, zooming in on your face and everyone else watching. Some were nervous that they might panic, splutter or forget the question they wanted to ask. For many students, therefore, the interactivity was very important – as long as it was someone else asking the questions!"

interface	In electronics, a shared boundary between two components with a defined form of signals at the boundary. A well-known example is the *V.24* interface between a microcomputer and a modem. See also *user interface*.
Internet	The inter-operating collection of academic computer networks which link most universities in most developed countries into a world-wide 'web' or 'metanetwork'. The Internet uses many digital circuits (at 64 kbit/s and 2 Mbit/s in Europe) but does not use ISDN technology – yet.
ISDN	Integrated Services Digital Network.
ISDN-2	Basic Rate ISDN (2 channels).

ISDN-30   Primary Rate ISDN (30 channels available to the user).

ISDN advantages over PSTN
                           These include the following:

i)   Error-free connections.

ii)   Fast call set-up.

iii)   Higher data rate than modems.

iv)   Predictable performance.

ISDN applications The following list should be taken as general guidance only:

i)   LAN to LAN interconnection, especially for backup and occasional extra capacity.

ii)   Transfer of large files such as are found in medical and publishing applications.

iii)   Video conferencing and group work generally.

iv)   General replacement and upgrade of modem-based applications.

v)   Access to multimedia databases (for home shopping, etc).

vi)   High-quality facsimile.

vii)   Electronic mail of multimedia messages.

ISDN over LANs   The current generation of LANs (Ethernet and Token Ring) cannot adequately handle real-time channels and thus cannot handle ISDN. This makes it hard to provide ISDN access throughout an organisation since network managers are reluctant to install new cabling. Originally the PTTs and the manufacturers of PBXs hoped that ISDN would be used for the LAN but this is now regarded as unrealistic. Even FDDI, a newer generation of LAN, does not adequately handle ISDN unless extra features are added.

ISO   International Standards Organisation. The main agency of the United Nations concerned with international standardisation (but see *ITU*).

ISPBX   Integrated Services Private Branch Exchange. A telephone switchboard oriented to switching ISDN connections within an organisation and from the organisation to the outside world.

ITU   International Telecommunications Union. An agency of the United Nations with a brief to foster international co-operation in telecommunications. It has three permanent committees of which the most important to us is the *CCITT*.

J

JANET   Joint Academic NETwork. A data communications network linking all universities and many other tertiary education institutions in the UK, which is connected to similar networks in other European countries and the US. JANET is a component network of the Internet.

JPEG   Joint Photographic Expert Group. A group who have defined a standard for still picture compression. Various vendors have applied it also to moving pictures, but better methods for these are now available – see *MPEG*.

K

kb   A unit of measurement of storage – kilobytes. 1 kb is 1024 bytes. (A byte is 8 bits.) Examples:

i)   An old-fashioned PC had a memory of 640 kb.

ii)   A colour super VGA image (640 by 480) is 320 kb.

kbit/s   A unit of measurement of data rate – kilobits per second. 1 kbit/s is 1000 bits per second. Examples:

i)   An old-fashioned modem has a speed of 0.3 kbit/s (300 bit/s).

ii)   A modern mass-market modem has a speed of 2.4 or 9.6 kbit/s.

iii)   The fastest modem under development runs at 28.8 kbit/s.

iv)   ISDN runs at multiples of 64 kbit/s.

v)   The aggregate bit rate of an ISDN-2 link is 144 kbit/s.

vi)   Excellent quality videoconferencing is available at 384 kbit/s.

L

LAN

See *Local Area Network*.

leased line

A dedicated circuit which runs from one point to another. Most parts of most wide area networks are built out of leased lines. Leased lines can be analogue or digital but nowadays in most countries are digital. The typical 'entry-level' speed for a European leased line is 64 kbit/s – the speed of ISDN. The next main stopping point (in Europe) is 2 Mbit/s – the speed of ISDN Primary Rate. These equalities are no coincidences – ISDN in many ways is the extension of the digital leased line concept to a dial-up methodology. These equalities also should make it easy for network planners to integrate leased lines with ISDN – but few of them are yet doing that.

Leased lines are paid for on a monthly rental, irrespective of the amount of traffic on the line or the amount of time it is used. Thus the tariffing of leased lines is quite unlike that for PSTN, ISDN or X.25 services. By and large, network planners prefer predictable costs and this is one reason (there are others) why they are continuing to favour leased lines even in the ISDN era.

light pen

A pen-like device that can be used to 'write' directly on a computer screen. It can be used to write freehand on the screen, to interact with a menu/windowing system on the screen and to enter and edit graphics. Light pens avoid the problems of hand-eye co-ordination that occur with graphics tablets (where one is drawing on one surface but what one draws is appearing on another). On the other hand, it is hard to write well with a light pen, especially on a vertical screen. Even if the screen is tilted so that the writing angle is more horizontal, it is still hard because the surface feel is so different.

LIVE-NET

This is the name of two distinct systems:

i) The University of London video network. (See Chapter 13 by Beckwith.)

ii) An Australian satellite trial starting in 1990 of two-way video, two-way audio at 384 kbit/s at four sites, originated by Karratha College and involving its two branches in the north-west of Western Australia and the Central Metropolitan College of TAFE in Perth.

local area network A method of linking computers within a restricted geographical area such as a building or campus Examples include Ethernet, Token Ring and LocalTalk.

lossless A method of data compression which does not lose any information in the signal.

lurker In a computer conferencing system, someone whose activity in the system is confined to reading messages. A lurker does not reply to messages or create any new messages. It is hard to detect the lurker's presence, hence the name.

M

Macintosh (Often abbreviated to Mac.) A series of computers developed by Apple with a very user-friendly and influential graphical user interface. Macintosh computers are much less prevalent than PCs but very popular with those that have them – they tend to be concentrated in universities and in specific applications areas such as desktop publishing, presentation graphics and multimedia. There are rather few ISDN solutions for Macintosh computers but at least two vendors make plug-in ISDN cards (for the Nubus slots) and Terminal Adaptors can of course be used. Several interesting desktop conferencing systems are available for the Macintosh.

MAN See *Metropolitan Area Network*.

Mb A unit of measurement of storage – megabyte. 1 Mb is 1024 bytes. (A byte is 8 bits.) Examples:

i) The memory in a typical PC-compatible running under Windows is 2 or 4 Mb.

ii) The storage on a floppy disc is 1.4 Mb.

iii) A small-capacity hard disc drive contains 40 Mb.

iv) A large hard disc as used on a powerful business micro could have 240 Mb of storage.

MBA   Masters Degree in Business Administration. Many hypothetical tele-educational case studies use the example of MBA courses; and some MBA courses do use tele-education techniques. It is often thought that (a) MBA students will be more able to afford the kind of equipment needed to provide tele-education and (b) the high fees charged on MBA courses make it easier for universities to invest in tele-education.

Mbit/s   A unit of measurement of data rate – megabits per second. 1 Mbit/s is 1,000,000 bits per second. Examples:

i) An Ethernet LAN runs at 10 Mbit/s.

ii) A Token Ring LAN runs at 4 or 16 Mbit/s.

iii) The new FDDI standard for LANs runs at 100 Mbit/s. There are also developments of Ethernet over twisted copper wire which run at this speed.

iv) ISDN Primary Rate runs at 2 Mbit/s.

v) Until recent years videoconferencing codecs needed 2 Mbit/s. Now they can run at 0. 384 or even 0.128 Mbit/s.

vi) Beyond 2 Mbit/s the next stage in the PTT digital hierarchy is 34 Mbit/s. This is seen in the modern view as the threshold of broadband.

media substitution A planning methodology for educational networks which assumes that one *medium* can more or less be replaced by certain others. This allows one to calculate the costs of various different approaches to solving an education or training requirement. Some educational experts are reluctant to accept this 'engineering' approach to planning the use of media in an organisation.

medium   This has two relevant meanings.

i) In data transmission, the 'material' over which data communication takes place. On an Ethernet, this would be coaxial cable, fibre or twisted pair. On a VSAT service, the medium would be free space.

ii) In educational technology, the medium is what carries the message. Media include television, print, electronic mail, videoconferencing, etc. The medium in an educational technology sense would probably be called a "service" by network planners. There are long debates about how similar are various media (like television and video) and to what extent media can be substituted for each other.

meet-me bridge   A type of audio bridge that can be accessed by calling a certain telephone number. When the number answers, the caller is connected into the audioconference. Some such bridges can allow over 200 sites to be connected, though 10-site capability will suffice for most education and training applications. The principle can be applied to other sorts of bridges such as for videoconferencing.

mentoring   Informal guidance (such as of a research project) by a much more experienced (and usually older) person. It is thought that mentoring can be carried out by videoconferencing.

Metropolitan Area Network
    A computer network whose geographical scope is of the order of size of a city. A number of specific technological solutions, some derived from the cable TV industry are appropriate for such networks.

modem   An abbreviation for MODulator-dEModulator. A device which translates digital signals into analogue ones so that data can be carried over the PSTN. Modems operate at different speeds depending on the model – speeds are normally in the range 300 to 28800 bit/s.

MPEG   Motion Picture Experts Group. A group who have defined standards for motion video compression. The first iteration was oriented to use with CD-ROM.

MPEG-2   A newer version of MPEG targeted to use in compressed video broadcasting (such as over satellites – between 4 and 8 channels can be carried in the bandwidth previously used for just one analogue channel). Standards are supposed to be fixed by the end of 1993.

multimedia   Pertaining to the use of several media (text, diagram, image, audio or video), either separately or combined.

multiplexing   The combining of several channels of information into one channel.

multipoint   Pertaining to multiple teleconferencing locations such as those connected via a *bridge*.

N

narrowband   The opposite of broadband. Networks under 2 Mbit/s or 34 Mbit/s depending on one's point of view. See *broadband*.

network   A network consists of a number of information sources connected via a communications *medium* for the purposes of exchanging information. Types of network include *Local Area Networks* and *Wide Area Networks*.

NREN   National Research and Education Network. The proposed broadband successor to the Internet in the US.

NT1   Network Termination 1. The physical end of the ISDN line at the customer's premises.

NT2   Network Termination 2. The customer's access point.

NTE   Network Termination Equipment.

NTSC   National Television Systems Committee. The US standard for TV, which uses 30 frames per second with 525 lines per frame.

O

off-line   The opposite of *online*.

off-line reader   Software on a microcomputer which allows it to operate with a computer conferencing system by copying all the relevant material on to the microcomputer's storage and manipulating it there as far as possible (including composing messages and replying to other messages). This has the effect of minimising the amount of time that the microcomputer is online and thus minimising the telephone (or other network) charges.

Olympus
An experimental satellite launched by the European Space Agency and used by several projects for satellite television and videoconferencing.

online
A system is operating online if it is connected to a remote system. Thus a videoconferencing system is online if it is sending and receiving pictures from another system. A microcomputer is online if it is connected to a remote computer system (such as a database).

OSI
Open Systems Interconnection. The methodology which the standards bodies recommend for the protocols used by computer networks.

P

packet switching
A data transmission method that breaks a flow of data into smaller units, called packets. These are individually addressed and routed through a network. The public X.25 services are good examples of packet switching networks. Various developments of this basic methodology are at the heart of the communications systems proposed for Broadband ISDN.

PAL
Phase Alternate Line. The television standard used (with variants) in most countries of Europe (except in France and the others who use SECAM).

PC
Personal Computer. Normally used to mean IBM-compatible PC (that is, an MS-DOS machine).

PictureTel
One of the leading vendors of codecs.

pixels
PICture ELementS. The dots that go to make up a picture on a computer or television screen. A very common number of pixels on a screen is 640 horizontal by 480 vertical (Super VGA).

POTS

Plain Old Telephone Service. Often used in a derogatory way by proponents of ISDN or other modern networks. Synonym for PSTN.

Primary Rate ISDN The 30-channel version of ISDN, oriented to use by a company ISPBX.

PSTN

Public Switched Telephone Network. The usual telephone network. Often also called POTS.

PTT

Post, Telephone, and Telegraph Company. The government-authorised supplier of telecommunications services in a country (pre-deregulation).

Q

QuickTime

Apple's system for encoding text, graphics, still image, audio and video into a multimedia presentation. QuickTime incorporates its own video compression technology.

R

rate adaptation

This has two distinct but related meanings:

i)   The conversion between the European 64 kbit/s version of ISDN and the North American 56 kbit/s version.

ii)   The conversion in a Terminal Adaptor between the data rate on a V.24 serial interface (usually at most 9600 or 19200 bit/s) and the 64 kbit/s of ISDN.

real-time

Several authors (see for example Chapter 12) use real-time as a synonym for *synchronous*.

Red Book

The 1984 edition of the CCITT reports which, among many other things, control the "I-series" standards used for ISDN. Now superseded by the *Blue Book*.

resource-based learning

The use of educational resources (films, videos, textbooks, CBT software packages, computer databases, etc) to facilitate learning and to some extent to replace human tutors. Traditionally these resources were 'locked up' in a library or resource centre. It is now being proposed that students gain access to these resources either from CD-ROMs supplied to them or by online access, perhaps over ISDN. However, in order to make this concept work, a massive amount of digitisation has to take place and there are major problems such as copyright to be overcome.

RS232C   The US name for the serial interface standardised by CCITT as V.24.

S

scanner   A device for digitising anything in paper form (text, diagram, photographs). It is like a photocopier except that the image is not printed out but is stored in digital form, so that it can be transmitted over networks or manipulated by software. A facsimile system contains a scanner. Scanners are attached to many videoconferencing systems and are particularly useful on desktop conferencing and audiographic systems.

second-generation distance education
          The approach to distance education used by all of the existing distance education organisations for the majority of their activity. It is said by its detractors to be characterised in educational terms by a transmissive and behaviourist approach, in production terms by a mass-production 'Fordist' approach, in technological terms by use of print, one-way television and Computer Based Training, and in student terms by the fostering of an individualist competitive approach.

slow scan   A style of video communication where video images are transmitted at much lower rates than the 25 or 30 frames per second required for broadcast quality television.

Standards   Standards may be set by a number of organisations and in a number of ways. In computing, the best known but not necessarily the most important standards are those set by ISO. In telecommunications, the standards are set by CCITT (and some related bodies). In the area where computing and telecommunications overlap, CCITT and ISO work on the whole in harmony. ISO people are fond of dismissing the CCITT standards as merely 'recommendations' but since such recommendations apply to PTTs (the members of CCITT) they usually have more legal force than the ISO standards.

In many sectors the most important standards are often those set by an industry leader and/or a consortia of major companies. They are usually called de facto standards. Quite often ISO comes along later and produces an international standard based closely on the de facto standard. This seems to happen much less with CCITT.

SuperJANET   The emerging broadband upgrade to the existing JANET network for high-speed data traffic including video (and voice). Most UK universities will be connected to SuperJANET in the next two years.

supplementary services
                 Services additional but related to telephone service that can be provided over modern digital exchanges and the ISDN. They include call waiting indication, call diversion and calling line identification.

synchronous   This term is well-known in data transmission but is used differently in educational circles.

i)   In data transmission, synchronous transmission is a method of transmitting data in which data is sent in blocks, with each bit of the block sent at a predictable time.

ii)   In education, synchronous interaction is where the two parties need to be simultaneously present. Thus videoconferencing and audioconferencing are synchronous, whereas computer conferencing and electronic mail are regarded as *asynchronous* media. From Chapter 6 by Latchem: "Teaching by videoconferencing has been shown to be effective where interactivity and synchronous learning apply. It works best with teachers who like to use a conversational style and who prepare for, and encourage, voluntary contributions by all participants through tutorials, role plays, simulations, brainstorming, problem-solving and practical activities."

T

T1 A high-speed digital leased line service available in the US with a bit rate of 1.5 Mbit/s. Each T1 service can accommodate 24 voice channels (at 56 kbit/s). T1 is the US equivalent of the European 2 Mbit/s E1 leased line service.

TA Terminal Adaptor.

tariffs The charges for a service.

TCP/ IP Transmission Control Protocol/Internet Protocol. The protocol suite used on the US Internet and on many European academic networks, as well as on many local area networks.

TCP/IP over ISDN Some experiments have been carried out on using the TCP/IP protocol over ISDN. (This would allow better integration of ISDN with the Internet.)

TDM Time Division Multiplexing.

TE Terminal Equipment. The phrase in the ISDN literature to denote equipment such as telephone handsets, facsimile machines and computers.

    i)   TE1 is the grade of terminal equipment that can connect directly to ISDN

    ii) TE2 is the grade of terminal equipment that requires a Terminal Adaptor to interface it to ISDN.

tele-education A subspecies of distance education, tele-education is the use of telematic networks for educational purposes.

teleconferencing See *conferencing*.

telematic A word derived from the French "telematique" meaning "related to the use of communication networks".

telematic network A network carrying speech, data, video or any combination of these. ISDN is an example.

telepresence

In teleconferencing situations, the use of communications technology to provide each user at a site with the feeling that the users at other sites are physically present. Obviously having two-way full motion video on large screens at each site is a powerful aid to telepresence, but the term is especially applied to careful use of non-video techniques. For example, research shows that close attention to audio quality and use of *light pens* (whatever is drawn) are powerful aids to telepresence. Even in asynchronous media such as *computer conferencing*, there is scope for increasing the amount of telepresence, for example by real-time notification when users log in and log out, and by maintenance of lists of who has read what messages.

teletex

An enhanced version of telex, based on a concept of communicating word processors, which was proposed by the CCITT and deployed for a while in some countries. It was expensive and hard to integrate with modern PCs and desktop publishing systems. Teletex suffered from competition from mainframe-based electronic mail systems, including those offered by PTTs as rivals to teletex! However, the general approach of teletex is becoming popular again as part of the move away from mainframes to client-server systems, but it will not be the CCITT version of the system that is used.

telewriting

Strictly speaking, the transmission of handwriting in real time from one system to another. More generally, a system which offers not just the transmission of handwriting but also other features such as transmission of pre-stored graphics and creation of simple graphics in a graphical shared screen environment. Telewriting often appears as a component of other systems.

Well-known telewriting systems include the currently available Optel system and the Cyclops system which flourished in the 1980s.

Terminal Adaptor

An ISDN device which interfaces traditional telecommunications equipment (such as telephones) and/or data communications equipment (usually with a V.24 serial interface) to the ISDN.

third-generation distance education

To quote from Chapter 11 by Kaye: "a move away from transmissive and behaviourist models of education and training towards constructivist models, which place more emphasis on inter-personal collaboration, social networking, peer exchange and group activity in the learning process, and which recognise that learning cannot be considered in isolation from contextual and social factors. This movement is necessarily accompanied by changes in the traditional roles of both learners and tutors: learners are expected to take a more active and constructive role, contributing from their own experience and knowledge as appropriate; tutors are expected to be facilitators and resource people, as well as course planners and knowledge brokers."

There are no real examples as yet of third-generation distance education systems operating on a large or comprehensive basis. Nevertheless, theorists have already suggested that the appropriate technologies for third-generation distance education systems will be videoconferencing, computer conferencing and database access (to compendia of course materials).

A related concept to third-generation distance education systems is the post-Fordist model, which places reliance on just-in-time creation of course material by small task groups.

Time Division Multiplexing

The key technology used in the backbone trunk network for telephony. It works by combining in rapid sequence samples of many channels into one trunk channel.

transmissive

A pejorative phrase associated with the concept of second-generation distance education. Transmissive models of education assume no feedback and use one-way media such as print and television.

Tutored Video Instruction

A way of combining low-budget video with a tutor who need not be a subject expert. Rather than watch the video by themselves, the students watch the video under the direction of a tutor who can help with difficulties and organise assessment and feedback.

TVI upgrades easily into a videoconferencing scenario.

TVI   Tutored Video Instruction.

TVRO   TeleVision Receive Only. A satellite dish and receiver combination to receive television from a Direct Broadcast Satellite or other television relay satellite. The less powerful the satellite, the larger (and thus more expensive) the dish and the more sophisticated (and thus expensive) the receiver.

twisted pair   A pair of copper wires twisted together, such as is used for telephone wiring. A somewhat more sophisticated version of this is used in many Local Area Networks.

U

uplink   A satellite earth station which transmits signals (such as compressed video) to a satellite. The opposite of *downlink*.

user-friendly   A system is called user-friendly if it is able to be learnt by the user with a minimum of training. It is important to remember that (a) user-friendly is a relative term and that (b) user-friendliness of a system depends on the user and what other systems the user knows. One of the reasons why windows-based software systems seem to be getting more user-friendly is that the more accustomed the user becomes to windowing the more user-friendly a new windows-based system seems to be.

user interface   The way in which a system such as a piece of software presents itself to the user. Modern software systems present themselves with a *graphical users interface*, that is, a windowing user interface similar to that found on the Apple Macintosh microcomputer.

V

V series   The series of CCITT Recommendations governing data transmission over telephone lines. The following entries list the main V recommendations of interest to users in the ISDN world.

V.22 bis	The CCITT standard for 2400 bit/s modems over dial-up lines. These modems are the ones most commonly purchased today for dial-up applications. Costs are in the order of £200 for cheaper models.
V.24	Confusingly, this is not a standard for modems, but for the serial connection between a modem and a computer. Traditionally, the full V.24 standard requires a 25-pin connector. In practice, 9-pin connectors are often used. A minimal subset of V.24 requires only 3 pins (transmit, receive and ground) but this minimal subset leaves out important functions for controlling the circuits.
V.32 bis	The CCITT standard for 14400 bit/s modems over dial-up lines. This is the fastest modem type which is in widespread use. Costs range from about £350 upwards.
V.32terbo	Despite its name, this is not a CCITT standard. It is a proposal from a group of modem manufacturers for a 19.2 kbit/s modem standard so that fast modems can be manufactured in advance of the time that *V.fast* modems are expected on the market.
V.42	The CCITT standard for error correction for modems. Error correction is needed because phone lines are noisy and so introduce errors into the data.
V.42 bis	The CCITT standard for data compression for modems. A compression ratio of up to 4:1 can be obtained.
V.fast	The proposed CCITT standard for 28800 bit/s modems (twice the speed of V.32 bis) over dial-up lines. Combined with V.42 bis, this promises an effective data rate of 112 kbit/s. Enthusiasts for V.fast then say "This is nearly twice as fast as ISDN!".
	Of course, data compression can also be applied on ISDN circuits, and the V.fast modems are unlikely to be able to work at full speed except on a small number of good-quality telephone lines.
VCR	Video Cassette Recorder.
VGA	Virtual Graphics Array. A commonly-occurring standard for the pixel and colour resolution of computer graphics displays.

VHS

Video Home System. The particular method of encoding video on a VCR which has gained mass commercial acceptance despite being allegedly less sophisticated than some rivals. Those arguing against D2-MAC often refer to VHS as an exemplar of the phenomenon that the market decides, not the technology (nor governments).

video disc

A 12" disc rather like a long-playing record that can store full motion video, audio and data. Modern video disc players can operate under computer control. Video disc was quite popular some years ago for delivering *interactive* education, but the technology was expensive and non-digital and is gradually being superseded in many applications by compressed video technologies based on CD-ROM.

videoconferencing

From Chapter 2 by Mason: "Videoconferencing is a confusing term in that it encompasses on the one hand, live video-lecturing to large audiences, and on the other, desktop conferencing point-to-point. The majority of large-scale applications currently, are satellite based – one way video, two way audio. Students at a remote site, or sites, see and hear the lecturer on several monitors positioned around the room. Using a telephone, they can call in questions live to the remote lecturer. ISDN promises to make two way video equally as cost-effective but more interactive. With two way video, every site can hear and see and speak to any other site. ISDN also facilitates smaller scale, multi-site videoconferencing, which lends itself to seminars, peer interaction, demonstrations of equipment or procedures, or round-robin discussions and training sessions."

videotelephony

This is a term similar to videoconferencing. However the general emphasis is different: videoconferencing is more about the transmission of video between more than two sites with several people at each site, using large screens; videotelephony is more about the transmission of video between two sites with one person at each site using a product not unrelated to a conventional telephone but with a small screen. Note that the above discussion is about the ideal world – in practice the terms are not clearly differentiated, least of all in Chapter 7.

videotex

A CCITT term for a standardised teleservice concerned with the transmission of 'pages' of data over telephone lines and its display on modified TV equipment. The service has different names in different countries – Prestel in the UK, Minitel in France. Like several CCITT teleservices, videotex has not become popular, with the notable exception of France where Minitel is widespread.

At one point the CCITT had great hopes that ISDN would lead to a blossoming of videotex since it would allow an enhanced videotex with the transmission of computer graphics (alphageometric) and even photographs (alphaphotographic) for applications such as shopping at home. However, the basic videotex user interface paradigm, as well as the technology, is now obsolete (being overtaken by PC-based windowing approaches and client-server systems respectively); so that in our view it is too late for ISDN to rescue videotex in its current form.

virtual reality

The creation using high-powered workstations and digital video of a naturalistic simulation of the real world. Denizens of virtual reality normally have to wear special helmets and 'data gloves'. The bandwidth and computer power required to generate virtual reality is currently very high; thus if the technology is to be available over networks it will be in the era of *Broadband ISDN*. Notwithstanding this, some more 'far-sighted' educators are already polishing up their grant proposals to look at applications of virtual reality to education and training. After all, simpler versions of virtual reality have been used in aircraft flight simulators for many years.

VSAT

Very Small Aperture Terminal. A satellite system (for voice, data or video) where the satellite dish is less than about 3 metres in diameter. VSATs are a type of downlink, but may also have uplink capability. TVROs are a restricted type of VSAT.

W

WAN

Wide Area Network.

Wide Area Network

An organisation network which covers a wider area than a LAN, perhaps even the whole world. In a modern organisation, a Wide Area Network is normally implemented mostly by using digital circuits leased from the PTTs. Other components of WANs include X.25 services and VSAT networks. Emerging technologies for WANs include Frame Relay (an optimised form of X.25) and ISDN.

workstation

Traditionally, a high-powered desktop computer (often using the Unix operating system) used for compute-intensive applications such as Computer Aided Design. In practice, used to mean any high-powered microcomputer – usually higher-powered than the one on your desk at present.

X

X series

The series of CCITT Recommendations governing data transmission over public data networks. There are many X recommendations, several very well-known such as X.25 and X.400.

However, few of the X recommendations directly concern the ISDN world and none of these are well known, even to many data communications experts.

X.25

The CCITT standard for public packet switch networks. Every developed country has a public packet switch network. The names vary with the country – the French system is called Transpac.

X.32

A CCITT standard for the use of dial-up connections (telephone or ISDN) by X.25 systems such as microcomputers. This standard is our tip for one that will become much better known in the ISDN world.

X.400

The CCITT standard for electronic mail systems. Several countries have implemented a public X.400 service – such as Gold-400 in the UK. Such services are normally connected to the *Internet*.

Y

Y-NET

An experimental X.400 service for European research.

The spectrum of kilobit rates

0.05	The speed of telex. However, since the telex character set uses only 5 bits, this is faster than you think – a little – all of 10 characters per second.
0.075	The speed of the V.23 modem (as widely used in videotex services) in the direction terminal to central computer. This corresponds to 7.5 characters per second and was alleged to be fast enough to cope with normal user typing. Clearly they never tried it out with trained typists!
0.11	The speed of the modems that used to be used with teletypewriters.
0.3	The speed of a V.21 bis modem as used for many years, but dying out fast.
1.2	The speed of a V.23 modem (central computer to terminal) as used in the majority of videotex services.
2.4	The speed of a V.22 bis modem. These are very widely used. Also tends to be the maximum speed at which one can access over the PSTN the majority of dial-up packet switching networks.
9.6	The speed of a V.32 modem. Some PSTN dial-up packet switching services (such as Transpac) now support access at this speed. Also the speed of a facsimile modem (the widely used Group 3). Also about as fast as one will need for accessing text-based services using a command-based interface.
14.4	The speed of a V.32 bis modem.
16	The speed of the D-channel (in Basic Rate ISDN).
19.2	A recognised CCITT speed for the V.24 interface. Several Terminal Adaptors can interface this speed, but no higher V.24 speed, to ISDN. Also note that many PCs and some Macintosh computers cannot reliably drive their serial interfaces at this speed!
28.8	The speed of the fastest modem under development which conforms to a CCITT standard.

56	The speed of the lowest grade of digital leased circuit in North America. Also the speed of the Accunet precursor of ISDN. Also about the maximum speed that can be obtained with a V.32 bis modem using V.42 bis data compression. (But data compression can be used with ISDN also.)
64	The speed of the ISDN B-channel (in both Europe and North America).
128	The lowest speed at which one can obtain reasonable quality with videoconferencing including audio.
144	The aggregate bit rate of Basic Rate ISDN (2 B + D).
384	Until recently, the lowest bit rate at which one could obtain good quality videoconferencing using CCITT standard codecs.
1544	The speed of the T1 grade of digital leased circuit widely used in North America (for example, for linking LANs).
2048	The speed of the E1 digital leased line service in Europe (corresponding to the North American T1). This is about the highest speed of leased line service generally available in Europe, even in the more advanced countries. Until recently, videoconferencing systems required this speed.
4000	The speed of the lower grade of Token Ring LAN.
8448	(8 Mbit/s.) The speed at the second order of the PTT multiplexing hierarchy.
10000	The speed of Ethernet.
16000	The speed of the higher grade of Token Ring LAN.
34368	(34 Mbit/s.) The speed at the third order of the PTT multiplexing hierarchy. Regarded by advanced thinkers as the current threshold of broadband. Also being considered as a low-speed B-ISDN channel.
44736	(Usually regarded as 45 Mbit/s.) The speed at the third order of the North American multiplexing hierarchy. Also the speed of T3 digital leased lines used by Fortune 100 companies to build their corporate Wide Area Networks.
100000	The speed of an FDDI LAN.

139264	(Usually regarded as 140 Mbit/s.) The speed at the fourth order of the PTT multiplexing hierarchy.
155520	(155 Mbit/s.) The lower of the proposed B-ISDN data rates.
565148	(565 Mbit/s.) The speed at the fifth (and currently last) order of the PTT multiplexing hierarchy. Note that this is over 10 million times the speed of telex.
622080	(622 Mbit/s.) The higher of the proposed B-ISDN data rates.
1000000	(1000 Mbit/s – 1 gigabit per second, Gbit/s.) Here we enter the realm of experimental fibre optic networks. BT have carried out a trial of a 2400 Mbit/s network.

Contributors

Chapter 1

Gabriel Jacobs is at the School of European Languages, University College of Swansea, Singleton Park, Swansea SH2 8PP, Great Britain.

Chapter 2

Dr. Robin Mason is Lecturer in IET (The Institute of Educational Technology) at the Open University, Walton Hall, Milton Keynes MK7 6AA, Bucks, Great Britain.

Chapter 3

Dr. Paul Bacsich is Senior Project Manager, Information Technology, and Head of the Electronic Media Research Group, at the Open University, Walton Hall, Milton Keynes MK7 6AA, Bucks, Great Britain.

Chapter 4

Edward G. Borbely is Director, Off-Campus Education, Columbia University, School of Engineering and Applied Science, 410 SCEPSR, New York, NY 10027, USA.

Chapter 5

Prof Patrick Purcell and Gerard Parr are at the University of Ulster MF141, Rock Road, Londonderry, BT48 7JL, Northern Ireland.

Chapter 6

Colin Latchem is Associate Professor, Head of Teaching/Learning Group, Curtin University of Technology, PO Box U 1987, Western Australia 6001, Australia. John Mitchell was formerly at South Australia Department of Employment and Technical and Further Education, Adelaide, South

Australia. (He is now Managing Director of J. G. Mitchell and Associates.) Roger Atkinson is at Murdoch University, Perth, Western Australia.

Chapter 7

Tove Kristiansen is at the Research Department of Norwegian Telecom, Postboks 83, 2007 Kjeller, Norway.

Chapter 8

Dr James Lange is at PictureTel, 258 Bath Rd, Slough, Berkshire SL1 4DX, Great Britain.

Chapter 9

Jean-Louis Lafon is at ENIC, 6 Rue des Techniques, 59658 Villeneuve – d'Ascq Cedex, France.

Chapter 10

Niki Davis is at the School of Education, Exeter University, Heavitree Rd, Exeter, Devon, EX1 2LU, Great Britain.

Chapter 11

Tony Kaye is Senior Lecturer in IET (The Institute of Educational Technology) at the Open University, Walton Hall, Milton Keynes MK7 6AA, Bucks, Great Britain.

Chapter 12

Peter Zorkoczy is at EPOS International, Zurcher Strasse 111, CH-8640 Rapperswil, Switzerland.

Chapter 13

Richard Beckwith is at the Imperial College of Science, Technology and Medicine, Prince Consort Road, London SW7 2BP, Great Britain. (At the time of writing this paper he was Director of LIVE-NET, University of London, Senate House, Malet Street, London WC1E 7HU.)

Index

Printed in the USA
CPSIA information can be obtained
at www.ICGtesting.com
JSHW011509221024
72173JS00005B/1250